T0305996

Materials Science
and Technology for
Nonvolatile Memories

MATERIALS RESEARCH SOCIETY
SYMPOSIUM PROCEEDINGS VOLUME 1071

Materials Science and Technology for Nonvolatile Memories

Symposium held March 24–27, 2008, San Francisco, California, U.S.A.

EDITORS:

Dirk J. Wouters
IMEC
Leuven, Belgium

Seungbum Hong
Argonne National Laboratory
Materials Science Division
Argonne, Illinois, U.S.A.

Steven Soss
Numonyx Corporation
California Technology Center
Santa Clara, California, U.S.A.

Orlando Auciello
Argonne National Laboratory
Materials Science Division and
Center for Nanoscale Materials
Argonne, Illinois, U.S.A.

Materials Research Society
Warrendale, Pennsylvania

CAMBRIDGE
UNIVERSITY PRESS

University Printing House, Cambridge CB2 8BS, United Kingdom

One Liberty Plaza, 20th Floor, New York, NY 10006, USA

477 Williamstown Road, Port Melbourne, VIC 3207, Australia

314-321, 3rd Floor, Plot 3, Splendor Forum, Jasola District Centre, New Delhi - 110025, India

79 Anson Road, #06-04/06, Singapore 079906

Cambridge University Press is part of the University of Cambridge.

It furthers the University's mission by disseminating knowledge in the pursuit of education, learning and research at the highest international levels of excellence.

www.cambridge.org
Information on this title: www.cambridge.org/9781605110417

Materials Research Society
506 Keystone Drive, Warrendale, PA 15086
http://www.mrs.org

© Materials Research Society 2008

This publication is in copyright. Subject to statutory exception and to the provisions of relevant collective licensing agreements, no reproduction of any part may take place without the written permission of Cambridge University Press.

This publication has been registered with Copyright Clearance Center, Inc. For further information please contact the Copyright Clearance Center, Salem, Massachusetts.

First published 2008
First paperback edition 2012

Single article reprints from this publication are available through University Microfilms Inc., 300 North Zeeb Road, Ann Arbor, MI 48106

CODEN: MRSPDH

A catalogue record for this publication is available from the British Library

ISBN 978-1-605-11041-7 Hardback
ISBN 978-1-107-40853-1 Paperback

Cambridge University Press has no responsibility for the persistence or accuracy of URLs for external or third-party internet websites referred to in this publication, and does not guarantee that any content on such websites is, or will remain, accurate or appropriate.

CONTENTS

*Invited Paper

v

OXIDE RESISTIVE SWITCHING MEMORY

ORGANIC RESISTIVE SWITCHING MEMORY

NANOPARTICLE-BASED ORGANIC MEMORY

*Invited Paper

vii

FUTURE EXPLORATIVE MEMORY
CONCEPTS

PREFACE

The symposium "Materials Science and Technology for Nonvolatile Memories," held March 24–27 at the 2008 MRS Spring Meeting in San Francisco, California, was a follow-up of the 2006 Spring Meeting symposium on "Science and Technology of Nonvolatile Memories" (*Mat. Res. Soc. Symp. Proc. Vol. 933E*) and also the 2004 Fall Meeting symposium "Materials and Processes for Nonvolatile Memories" (*Mat. Res. Soc. Symp. Proc. Vol. 830*), and the 2007 Spring Meeting symposium "Materials and Processes for Nonvolatile Memories II" (*Mat. Res. Soc. Symp. Proc. Vol. 997*), and was the fourth symposium in a series of MRS Meeting symposia on nonvolatile memories.

The strong attendance and large paper submission (in total 51 oral and 60 poster contributions were presented in 9 sessions, together with 10 invited talks), indicate the continuing strong international interest and research effort in the field of emerging new nonvolatile memory materials. Main areas of research are advanced Flash and resistive switching RAM including cross-point and organic memories, while ferroelectric-ferromagnetic and ferroic materials also remain of interest (it must be noted that phase change material was mostly covered in a separate symposium at the same conference).

The selected papers in this proceedings volume have been categorized into seven chapters. The first chapter *Advanced Flash Memory* deals with solutions for scaled Flash memory, including the use of new high-k layers and nanocrystals. Resistive switching concepts are discussed in the *Oxide Resistive Switching Memory* as well as in the *Organic Resistive Switching Memory* chapters. More research on polymer memories can be found in the chapters on *Nanoparticle-based Organic Memory* and *Organic Ferroelectric Memory*. The final two chapters deal with *New Phase Change Memory and Deposition Methods,* and *Future Explorative Memory Concepts*, the latter including piezoelectric, ferroelectric and ferromagnetic concepts.

A highly successful one-day tutorial was conducted, including tutorials on oxide resistive random access memories (OXRRAM), magnetic random access memories (MRAM), probe storage and phase change memories (PCM). The tutorials were very well attended (60–100 people) by scientists, postdoctorals and students, providing an opportunity for detailed discussion on the addressed topics.

With contributions from university, research centers and industry, the papers from this symposium proceedings reflect the recent evolutions in material technology and the understanding in these different fields.

Dirk J. Wouters
Seungbum Hong
Steven Soss
Orlando Auciello

June 2008

ACKNOWLEDGMENTS

The Symposium Organizers would like to thank the tutorial speakers as well as the invited speakers who contributed to the success of this symposium:

Tutorial Speakers:
Greg Atwood (Intel Corp., Santa Clara, CA), Claude Chappert (CNRS, France), Yasuo Cho (Tohoku Univ., Japan), and Rainer Waser (RWTH Aachen, Germany)

Invited Speakers:
R. Ramesh (Univ. of California-Berkeley, CA), Jun Hayakawa (Hitachi Advanced Research Laboratory, Tokyo, Japan), Edwin C. Kan (Cornell University, Ithaca, NY), Jan Van Houdt (IMEC, Leuven, Belgium), Armin Knoll (IBM Research GmbH, Ruschlikon, Switzerland), Hyoungsoo Ko (Samsung Advanced Institute of Technology, Gyeonggi-do, South Korea), Myoungjae Lee (Samsung Advanced Institute of Technology, Gyeonggi-do, South Korea), Robert Mueller (IMEC, Leuven, Belgium), Yan Yang (UCLA, Los Angeles, CA), and Carlos A. Paz de Aroujo, (Symetrix Corp., Colorado Springs, CO).

We also wish to thank the following organizations for their financial support of this symposium:

Applied Materials Inc.
Intel Corporation
Seagate Technology
Symetrix Corp., Colorado Springs

MATERIALS RESEARCH SOCIETY SYMPOSIUM PROCEEDINGS

MATERIALS RESEARCH SOCIETY SYMPOSIUM PROCEEDINGS

Volume 1087E —Crystal-Shape Control and Shape-Dependent Properties—Methods, Mechanism, Theory and Simulation, K-S. Choi, A.S. Barnard, D.J. Srolovitz, H. Xu, 2008, ISBN 978-1-60511-057-8

Volume 1088E —Advances and Applications of Surface Electron Microscopy, D.L. Adler, E. Bauer, G.L. Kellogg, A. Scholl, 2008, ISBN 978-1-60511-058-5

Volume 1089E —Focused Ion Beams for Materials Characterization and Micromachining, L. Holzer, M.D. Uchic, C. Volkert, A. Minor, 2008, ISBN 978-1-60511-059-2

Volume 1090E —Materials Structures—The Nabarro Legacy, P. Müllner, S. Sant, 2008, ISBN 978-1-60511-060-8

Volume 1091E —Conjugated Organic Materials—Synthesis, Structure, Device and Applications, Z. Bao, J. Locklin, W. You, J. Li, 2008, ISBN 978-1-60511-061-5

Volume 1092E —Signal Transduction Across the Biology-Technology Interface, K. Plaxco, T. Tarasow, M. Berggren, A. Dodabalapur, 2008, ISBN 978-1-60511-062-2

Volume 1093E —Designer Biointerfaces, E. Chaikof, A. Chilkoti, J. Elisseeff, J. Lahann, 2008, ISBN 978-1-60511-063-9

Volume 1094E —From Biological Materials to Biomimetic Material Synthesis, N. Kröger, R. Qiu, R. Naik, D. Kaplan, 2008, ISBN 978-1-60511-064-6

Volume 1095E —Responsive Biomaterials for Biomedical Applications, J. Cheng, A. Khademhosseini, H-Q. Mao, M. Stevens, C. Wang, 2008, ISBN 978-1-60511-065-3

Volume 1096E —Molecular Motors, Nanomachines and Active Nanostructures, H. Hess, A. Flood, H. Linke, A.J. Turberfield, 2008, ISBN 978-1-60511-066-0

Volume 1097E —Mechanical Behavior of Biological Materials and Biomaterials, J. Zhou, A.G. Checa, O.O. Popoola, E.D. Rekow, 2008, ISBN 978-1-60511-067-7

Volume 1098E —The Hydrogen Economy, A. Dillon, C. Moen, B. Choudhury, J. Keller, 2008, ISBN 978-1-60511-068-4

Volume 1099E —Heterostructures, Functionalization and Nanoscale Optimization in Superconductivity, T. Aytug, V. Maroni, B. Holzapfel, T. Kiss, X. Li, 2008, ISBN 978-1-60511-069-1

Volume 1100E —Materials Research for Electrical Energy Storage, J.B. Goodenough, H.D. Abruña, M.V. Buchanan, 2008, ISBN 978-1-60511-070-7

Volume 1101E —Light Management in Photovoltaic Devices—Theory and Practice, C. Ballif, R. Ellingson, M. Topic, M. Zeman, 2008, ISBN 978-1-60511-071-4

Volume 1102E —Energy Harvesting—From Fundamentals to Devices, H. Radousky, J. Holbery, B. O'Handley, N. Kioussis, 2008, ISBN 978-1-60511-072-1

Volume 1103E —Health and Environmental Impacts of Nanoscale Materials—Safety by Design, S. Tinkle, 2008, ISBN 978-1-60511-073-8

Volume 1104 — Actinides 2008—Basic Science, Applications and Technology, B. Chung, J. Thompson, D. Shuh, T. Albrecht-Schmitt, T. Gouder, 2008, ISBN 978-1-60511-074-5

Volume 1105E —The Role of Lifelong Education in Nanoscience and Engineering, D. Palma, L. Bell, R. Chang, R. Tomellini, 2008, ISBN 978-1-60511-075-2

Volume 1106E —The Business of Nanotechnology, L. Merhari, A. Gandhi, S. Giordani, L. Tsakalakos, C. Tsamis, 2008, ISBN 978-1-60511-076-9

Volume 1107 — Scientific Basis for Nuclear Waste Management XXXI, W.E. Lee, J.W. Roberts, N.C. Hyatt, R.W. Grimes, 2008, ISBN 978-1-60511-079-0

Prior Materials Research Society Symposium Proceedings available by contacting Materials Research Society

Advanced Flash Memory

Mater. Res. Soc. Symp. Proc. Vol. 1071 © 2008 Materials Research Society

1071-F02-01

Flash Memory Scaling: From Material Selection to Performance Improvement

Tuo-Hung Hou, Jaegoo Lee, Jonathan T. Shaw, and Edwin C. Kan
School of Electrical and Computer Engineering, Cornell University, Ithaca, NY, 14853

ABSTRACT

Below the 65-nm technology node, scaling of Flash memory, NAND, NOR or embedded, needs smart and heterogeneous integration of materials in the entire device structure. In addition to maintaining retention, in the order of importance, we need to continuously make functional density (bits/cm^2) higher, cycling endurance longer, program/erase (P/E) voltage lower (negated by the read disturbance, multi-level possibility and noise margin), and P/E time faster (helped by inserting SRAM buffer at system interface). From both theory and experiments, we will compare the advantages and disadvantages in various material choices in view of 3D electrostatics, quantum transport and CMOS process compatibility.

INTRODUCTION

Battery-powered portable electronics, such as mobile phones, MP3 players, digital cameras etc., have fuelled skyrocketing demand for nonvolatile Flash memory since late 1990's. The advance in technology is even more impressive. The Flash technology has demonstrated its outstanding scaling capability in the last decade. A two-fold increase in bit-density of NAND Flash has been realized every year for the past seven years [1]. Today 16-gagabit density with 50-nm design rule is in mass production. This trend far exceeds the projection of the Moore's law in logic integrated circuits. Therefore, Flash is arguably the present technology driver of the semiconductor industry. However, this great momentum, mainly relying on the straightforward geometrical shrinkage, has been expected to slow down for technology nodes of 40 nm and beyond due to several challenging roadblocks in device scaling [1-4]. First, the thickness of tunnel oxide is not easily scaleable in order for satisfactory charge retention, especially after many program/erase (P/E) cycles. The stress induced leakage current (SILC) gives rise to unacceptable statistical distribution in retention for a high-density memory array, which limits the thickness of tunnel oxide to be 7-8 nm [2, 3]. The non-scalable tunnel oxide deteriorates the short channel effects (SCE) and impedes further gate-length scaling. This is particularly severe in NOR-type Flash where the large drain voltage (> 3.2V) is necessary for hot-carrier programming. Second, the distance between adjacent float-gates (FGs) has become extremely narrow due to aggressive scaling. As a result, the cell-to-cell interference is no longer negligible. This in part can be mitigated by reducing the FG height and by utilizing a low-κ spacer between FGs. However, these inevitably hurt the coupling ratio (CR) necessary for decent P/E efficiency. In conventional designs, while the thickness of inter-poly oxide or so called control oxide is also reaching its scaling limit, the CR can still be engineered by the additional capacitance provided by FG sidewalls. The better immunity to the cell-to-cell interference by reducing the FG height is at the expense of the dwindling CR, and as a consequence even higher P/E voltage is required. P/E voltages are projected still at 15 V for NAND Flash until the end of roadmap in 2018 [5]. Higher P/E voltage leads to higher power dissipation and adversely affects the parallel writing process. It also adds tremendous overhead on power consumption and area of the peripheral circuit for both stand-alone and embedded memory [6]. Even more importantly, the endurance under many P/E cycles is deteriorated by the high field in the thin tunnel oxide. The resulting threshold voltage V_{th} shifting and SILC in short-retention bits are the key reliability concerns. Therefore, a fundamentally new approach to scale cell size without compromising memory performance is of great importance in Flash memory technology.

Meanwhile, there has been very active research on alternative nonvolatile memories that do not employ charge storage in FG. Among the most mature are ferroelectric random access memory (FRAM) [7], magnetoresistive random access memory (MRAM) [8], and phase-change random access memory (PRAM) [9]. Although enormous progress has been made, none of them have stood up to completely address the strict requirements for low-cost, high-density, and high-speed nonvolatile storage. FRAM relying on the charge polarization in small capacitors has limited scaling potential. Its destructive read is also undesirable. MRAM and PRAM are still under active investigation to realize an efficient P/E scheme compatible with the current drive capacity of scaled access transistors in the one-transistor-one-magnetic-tunnel-junction (1T1MTJ) and one-transistor-one resistor (1T1R) cells. In addition, any emerging technology has to be a cost-effective replacement, a daunting challenge considering the maturity of today's Flash technology as well as the prevalent implementation of multiple bits per cell. Therefore, it is safe to project that Flash memory will still be the main workhorse of the portable nonvolatile storage for many years to come [10-12]. The question is how we are able to extend its longevity by overcoming aforementioned scaling challenges before any viable alternative becoming a reality.

In an attempt to address this, in this paper we highlight the importance of smart and heterogeneous integration of materials throughout the entire device structure, including charge storage medium, tunnel oxide, control oxide, control gate, and sensing channel. From both theory and experiments, we will compare the advantages and disadvantages in various material choices in view of three-dimensional (3D) electrostatics, quantum transport and CMOS process compatibility. We will limit our discussion mainly on NAND-type Flash memory owing to its better scalability and dominate role in portable massive storage. However, many viewpoints presented here may apply to NOR-type Flash as well.

CHARGE STORAGE MEDIUM

Flash memory relies on the static-charge storage in an isolated FG. The conventional choice of material for FG has been n-doped poly-Si because of its process compatibility in the Si process. However, many aforementioned scaling challenges stem from the continuous poly-Si FG. Non-scaleable thickness of the tunnel oxide due to the poor immunity against SILC and significant cell-to-cell interference are two main inherent disadvantages. Although the industry has every reason to push the continuous FG to its limit, with recent demonstration in the 43nm-node technology [13], it is of little doubt that at some point discrete charge storage, which consists of multiple discrete FGs instead of a continuous one, has to be utilized in order to fundamentally resolve these issues [1, 2, 14]. The discreteness among FGs prevents complete loss of memory states through localized SILC, and greatly suppresses the FG-to-FG coupling. This enables both tunnel oxide and cell size scaling. Proposed device implementation is basically divided into two major categories, silicon-oxide-nitride-oxide-silicon (SONOS) or SONOS-type memories [15-25] and nanocrystal (NC) memories [26-42]. SONOS-type memories utilize natural traps in dielectrics while NC memories utilize semiconductor or metal NCs embedded inside dielectrics for charge storage. Here we are interested in the best option available to address the remaining roadblock, the high P/E voltage.

4

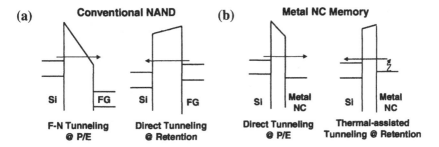

Figure 1 Energy band diagram representation at P/E and retention in the nonvolatile memory cells with (a) thick tunnel oxide and poly-Si FG and (b) thinner tunnel oxide and metal NC

In the present Flash memory, the ratio between retention time t_R and P/E time t_{PE} is about 10^{12}-10^{14}. In order to realize this tremendous ratio, field-asymmetric tunneling processes in the tunnel barrier have to be deliberately engineered between retention and P/E. The asymmetry in conventional Flash is most often provided by external P/E voltage. For example, in NAND Flash, the asymmetry between the Fowler-Nordheim (FN) tunneling under P/E and the direct tunneling (DT) during retention is exploited as illustrated in Fig. 1(a). However, this approach also limits the scalability of P/E voltage. Metal NC memory [32-37] has been proposed to enhance the tunneling asymmetry at low P/E voltage. The material-dependent FG work function of metal NCs provides additional band offset to the Si band edges of the channel. During retention, only a small portion of thermally excited charge in metal NCs is able to directly tunnel back to Si channel due to the Si forbidden bandgap. This greatly improves memory retention even with a thinner tunnel oxide. Meanwhile, the thinner tunnel oxide allows fast P/E operation through DT at low P/E voltage. The asymmetry between the DT under P/E and the thermal-assisted tunneling during retention as illustrated in Fig. 1(b) is fundamentally different from that in the conventional NAND. On the contrary, semiconductor NCs, such as Si, Ge, and SiGe NCs [26-30], provides little or none band offset to the channel. The quantum-size effect of semiconductor NCs further broadens bandgap larger than that in bulk Si. In metal NCs, this bandgap broadening is suppressed by the large density of states in metal for the size of metal NCs we are generally interested in [43]. Furthermore, previous studies suggested that charge retention in semiconductor NC memories is governed by interface traps surrounding NCs with deep energy level inside the Si bandgap [44, 45]. However, this mechanism is less reliable because the interface traps are subjected to many process variations such as the backend forming gas annealing, and there is no known method to reliably engineer deep-level traps. So are true for SONOS-type memories relying on bulk traps in dielectrics. Many studies have shown that retention at high temperature is problematic for SONOS with shallow-level traps [14, 20, 46]. In brief summary, the metal NC memory is a unique approach to further scale down the tunnel oxide without compromising retention. Therefore, the cell size scaling, low P/E voltage, and robust memory reliability may be realized simultaneously.

Electrostatics is another important consideration to achieve low P/E voltage. For better P/E efficiency, any potential drop on FGs has to be minimized especially with aggressive scaling on the thickness of tunnel and control oxide. In the conventional n-dope poly-Si FG, the poly depletion is present. In SONOS memories, the voltage drop on nitride is substantial because the nitride permittivity is only two times larger than SiO_2 and the thickness of nitride is comparable

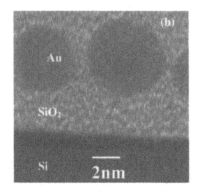

Figure 2 (a) A SEM plane-view image of Au NCs with area density of 4×10^{11}/cm^2, and (b) A STEM cross-sectional image of Au nanocrystals embedded in SiO$_2$ [34].

to oxide barriers. That is one of the reasons why higher-κ trap layers, such as Al$_2$O$_3$ [20], HfAlO [21], HfSiO [22], AlN [23], HfO$_2$ [24], and Ta$_2$O$_5$ [25], are more desirable. With the relatively higher Si permittivity and the small NC size, semiconductor NC memories seem to mitigate this concern. Nevertheless, as discussed in the later sections, the integration of high-κ dielectrics such as HfO$_2$ with κ = 20 as both the tunnel and control oxide makes the voltage drop on semiconductor NCs unavoidable. Therefore, metal NCs are the best option to eliminate the voltage drop with the orders of magnitude higher free electron concentration than the semiconductor counterparts. In addition, the above analysis is solely based on one-dimensional (1D) electrostatic approximation and too simplistic for NC memories because of the nature of the 3D spherical NCs and their two-dimensional (2D) placement. This is highlighted in the cross-sectional TEM and plain-view SEM in Fig. 2 [34]. Detail examination based on 3D electrostatics reveals the field-enhancement effects around NCs [47, 48]. For a typical design of metal NC memory, the potential drop in the tunnel oxide can be more than 40% higher than that in the continuous FG memory, resulting in great improvement on the P/E efficiency. This field enhancement is subject to not only geometrical parameters, many times being able to be solved only by numerical simulation, but also the choice of materials of NC and surrounding dielectric. Considering a simplified case when the top gate, the sensing channel, and other NCs are relatively far away and a NC with charge amount of Q stored is placed in a uniform field E_0, the analytical solution of the electric field intensity exists and can be expressed as [47, 48]:

$$E_r = E_0\left(1+\frac{2a^3}{r^3}\left(\frac{\varepsilon_{NC}-\varepsilon}{\varepsilon_{NC}+2\varepsilon}\right)\right)\cos\theta + \frac{\sum_i Q_i}{4\pi\varepsilon\, r^2} \qquad (1)$$

$$E_\theta = -E_0\left(1-\frac{a^3}{r^3}\left(\frac{\varepsilon_{NC}-\varepsilon}{\varepsilon_{NC}+2\varepsilon}\right)\right)\sin\theta \qquad (2)$$

where the origin of the spherical coordinate (r, θ) is at the center of the NC, a is the NC radius, ε_{NC} is the NC permittivity, ε is the dielectric permittivity, and θ is the angle between r and E_0. For a metal NC with infinite ε_{NC}, the field-enhancement term is reduced to $(a/r)^3$ even with high-κ dielectrics. On the other hand, for a Si NC with $\varepsilon_{NC} = 11.7$ embedded in SiO$_2$, the field-

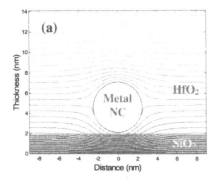 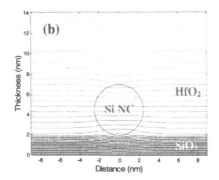

Figure 3 The cross-section view of the 3D electrostatic potential contours in NC memory unit cells with (a) a metal NC and (b) a Si NC. The NC diameter, the thickness of SiO_2 tunnel oxide, and the thickness of HfO_2 control oxide are 5 nm, 2 nm, and 7 nm, respectively. $V_G = 8$ V and no charge is stored in the NCs. The potential is monotonic from top to bottom, and the contour spacing is 0.2 V.

enhancement is merely $0.4 \times (a/r)^3$ and it gets less or even becomes negative when embedded in a high-κ matrix. The numerical simulation of 3D potential contours in a unit cell of metal and semiconductor NC memories is shown in Fig. 3 with a high-κ HfO_2 control oxide. The potential drop inside the Si NC and the electric field decrease around it are in strong contrast with the unit cell of metal NC memory. Therefore, metal NCs are preferable choice over semiconductor NCs as the integration with high-κ dielectrics is inevitable for future scaled memory devices [49]. Meanwhile, due to the infinitesimal physical size of traps, the $(a/r)^3$ term vanishes. Therefore, SONOS memories remain similar to the conventional continuous FG memory without additional field-enhancement from 3D electrostatics.

Despite aforementioned advantages, the discreetness of FGs also poses fundamental challenges in maintaining P/E efficiency. First, the charging energy arising from shrinking capacitance in the discrete FGs becomes substantial. Single-electron charging energy E_{CH} is the electrostatic energy required to store an additional electron in a small capacitor due to the Coulomb repulsion energy. It can be expressed as e^2/C where e is the elemental charge and C is the self-capacitance of the charge storage node from the 3D electrostatic calculation. C is a strong function of the NC size, and E_{CH} increases dramatically with the NC size scaling [50]. Therefore, in a typical design with a NC diameter of 5 nm, the maximum number of charges every NC can stably hold is around 10. In the SONOS-type memory, because of the infinitesimal size of traps, every trap can hold at most one charge. As a result, to warrant sufficient memory window and P/E efficiency, the NC or trap density has to be deliberately engineered to provide charge storage capability comparable to the conventional continuous FG. One interesting example is by stacking multiple layers of NC vertically to provide additional storage capacity [33, 39, 40]. Furthermore, small physical size of discrete FGs, also true for the extreme cases of traps, associates with small capture cross-sectional area σ during charge injection. This may adversely affect P/E efficiency. Lastly, the partial coverage of NCs over the surface of the Si channel results in less control on the channel potential, *i.e.* less memory window. A channel-control factor R between 0 and 1 is usually adopted in comparison with a continuous FG cell with R equal to 1 [48]. A smart design to boost R without increasing NC density will be further

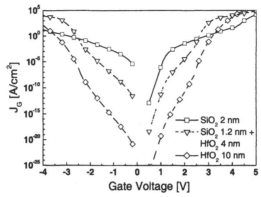

Figure 4 Calculated tunneling current from the WKB approximation for three tunnel oxide with the same 2-nm EOT.

discussed later by utilizing high-κ control oxide or small sensing channel. Overall NC with its moderate size provides low E_{CH}, large σ, and sufficient R, which greatly suppress adverse effects on P/E efficiency.

From the aspect of manufacturability, controlling tight V_{th} distribution at P/E states in a large memory array is very critical. It is a major drawback of the scaled NC memory cell as the fluctuation in the NC size and the NC number in each cell become substantial. However, it was projected that NC memory technology still has strong potential to scale beyond 65-nm node with current NC self assembly methods [14, 51, 52]. Recent efforts on ordered placement of NCs with controllable spacing of 3-15 nm [41, 42] may push the scaling limit even further. On the other hand, SONOS may provide better immunity to device variations owing to the large number density of traps. A heterogeneous NC/nitride stack may improve both V_{th} distribution and P/E efficiency for superior scalability [53, 54]. Both the semiconductor NC memory and the SONOS memory are fully compatible with the conventional Flash technology. They have been demonstrated for embedded applications to be fully compatible with CMOS, using even less masking steps compared with the embedded FG memory [16, 27]. High-κ trap layers and metal NCs are less compatible due to the concern of thermal stability and contamination. However, as high-κ dielectrics and metal gates become inevitable in the future Si technology, this may be less critical with better understanding and control on new material integration.

TUNNEL OXIDE

Tailoring the band structure of the tunnel barrier is another effective way to achieve significant tunneling asymmetry. High-κ dielectrics with lower electron / hole barriers are better field-sensitive tunnel barriers than SiO_2 [35]. In Fig. 4, tunneling current calculation based on Wentzel-Kramer-Brillouin (WKB) approximation [55] is shown for SiO_2 and HfO_2 with the same 2-nm effective oxide thickness (EOT). The WKB approximation of transmission probability T_{WKB} at the DT regime is expressed as:

$$T_{WKB} = \exp(-2\int_{0}^{t_{ox}} \sqrt{\Delta E - qF_{ox} \cdot x}\, dx) \qquad (3)$$

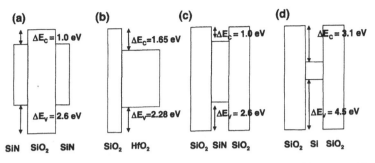

Figure 5 Energy band diagram representation of (a) crested barrier, (b) asymmetric layered barrier, (c) bandgap-engineered ONO, and (d) double tunnel junction.

where ΔE, F_{ox}, and t_{ox} are the dielectric / Si band offset, oxide electric field and oxide thickness, respectively. dT_{WKB}/dF_{ox} suggests T_{WKB} has stronger field dependence with smaller ΔE. Therefore, the current of HfO$_2$ has much stronger field dependence than that of SiO$_2$ at the DT regime. The lower transition voltage from DT to FN due to the lower ΔE further enhances the overall asymmetry. Composite tunnel barriers with multiple layers of dielectrics such as crested tunnel barriers [29, 56] and asymmetric layered barriers [49, 57, 58] as illustrated in Fig. 5 are designed by the same principle. In Fig. 4, a 1.2-nm SiO$_2$ + 4-nm HfO$_2$ with EOT of 2 nm exhibits similar field-sensitivity as a pure HfO$_2$ dielectric. The interfacial SiO$_2$ between high-κ dielectrics and the Si channel exists at many high-κ deposition processes, and also desirable to ease the severe mobility degradation caused by the remote phonon scattering [59] and reduce the interface traps that can affect cycling endurance.

The other class of field-sensitive tunnel barriers such as bandgap-engineered Oxide-Nitride-Oxide (ONO) [17], and double tunnel junction [19, 39] is also illustrated in Fig. 5. The structure consists of a small bandgap dielectric layer (SBL) sandwiched between two large bandgap dielectric layers (LBL). Resonant tunneling through the bound states at SBL is utilized to enhance the transmission probability at high field. However, this process is quenched at low field with bound state energy at SBL higher than the energy of injecting carriers. The only remaining transport is the DT current through the composite LBG/SBG/LBG, which is very low. Therefore, superior t_R / t_{PE} ratio at low P/E voltage has been demonstrated at highly scaled memory cells [39].

The employment of high-κ tunnel dielectrics is hampered by other disadvantages, such as mobility degradation in the channel and more importantly insufficient reliability caused by interface states D_{it} and dielectric traps. Transport mechanism of many high-κ dielectrics at low field is governed by the trap-assisted tunneling or interface-state assisted tunneling. Hence the large field-asymmetry estimated from an ideal high-κ dielectric is over optimistic. Furthermore, both natural and stress-induced traps in high-κ may degrade the cycling endurance and V_{th} distribution. However, through the advance of process technology, high-κ gate dielectrics have met strict reliability requirements for future CMOS [60]. Continuous P/E voltage scaling of Flash memory may eventually make reliable high-κ tunnel oxide feasible.

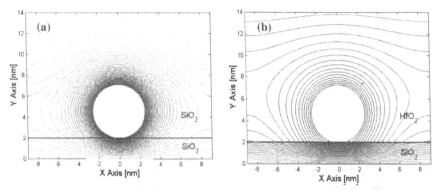

Figure 6 The cross-section view of the 3D electrostatic potential contours in the NC memory unit cell with (a) 7 nm SiO$_2$, and (b) 35 nm HfO$_2$ as the control oxide. Only part of the HfO$_2$ is shown in (b). The NC potential is set as -1 V while V_G = 0 V. The potential increases monotonically from the NC with the contour spacing of 25 mV [49].

CONTROL OXIDE & CONTROL GATE

Under P/E, the electric field at the control oxide increases significantly with the charge build-up at the FG, and so is the inter-poly leakage current. P/E saturation occurs when the inter-poly leakage current is comparable to the injecting current from the channel. As shown in Fig. 4, except under very high bias, high-κ dielectrics have substantially less leakage current than SiO$_2$ at the same EOT due to the physical thickness. Therefore, high-κ control oxide may be exploited to reduce the inter-poly leakage current and to increase CR simultaneously. This enables large memory window at lower P/E voltage or at higher P/E speed [61]. Combining with a metal electrode of high work function, the inter-poly current can be even further suppressed during erase [62].

In addition, spherical NCs are discretely placed on top of a 2D channel in NC memories. The coupling between NCs and the channel is subjected to 3D electrostatics. The detail of this coupling is important to determine the NC self capacitance, $i.e.$ E_{CH}. It is also important to determine the channel-control factor R. Smaller E_{CH} allows more charges being stably stored in NCs, and larger R provides wider memory window where $R = 1$ representing the upper limit of a continuous FG. Both are critical in optimizing memory P/E and retention characteristics. The cross-sections of the 3D potential contours in the NC unit cell with SiO$_2$ and HfO$_2$ control oxide are plotted in Fig. 6. The EOT remains the same for both stacks. It is obvious that the fringing field through HfO$_2$ to the Si channel is much stronger due to the higher permittivity of HfO$_2$. As a result, E_{CH} with HfO$_2$ is only a half of that with SiO$_2$, and R increases from 0.55 to 0.85 dramatically. This leads to the increase of t_R / t_{PE} ratio by more than four orders of magnitude [49].

Fermi-level pinning is known to shift the effective gate work functions of metal/high-κ and polysilicon/high-κ gate stacks substantially [63]. Similar effects have been found critical in NC memories integrated with high-κ control oxide [64]. The effective NC work function is not only a bulk property of the NC, but also governed by the interface with the surrounding dielectric due to the formation of interface dipoles. It has to be taken into account in engineering NC work function and design optimization of NC memories.

SENSING CHANNEL

Utilizing the field-effect transistor (FET) as a sensing channel sets Flash memory apart from other competitive nonvolatile technologies with very high sensitivity. However, the conventional planar FET on the bulk Si substrate is facing tremendous scaling challenges on its own. The short channel effects, line edge roughness, and substrate dopant fluctuation all make device variations intolerable in high-density memory array with aggressively scaled FETs. Flash memory based on novel FET structures aiming for better scalability has been under investigation [65-67]. In particular, FinFETs exhibits outstanding electrostatics from the 3D geometry of the control gate and the ultra-thin Si channel. Moreover, the fully-depleted ultra-thin channel warrants excellent immunity against the dopant fluctuation because V_{th} is controlled mainly by the gate work function instead of the substrate doping. The charge retention is also improved because of the floating body potential during retention. However, the same reason that prohibits FinFETs from replacing planar FETs in logic applications, high manufacturing cost, has to be first resolved.

Instead of the straight cell-size scaling, the other feasible way to increase the bit density per area is by stacking memory devices vertically [68-70]. This 3D staking approach has to be a cost-effective and low thermal-budget process without perturbing the characteristics of the underlying devices. The low-temperature thin-film transistor (TFT) is a promising candidate for 3D stacking, owing to its high maturity and low cost after active development for decades. High-performance control circuitry can be implemented on the Si substrate while the memory array is stacked multiple times vertically to reduce the chip size and bit-cost. Nevertheless, non-ideal subthreshold characteristics governed by high trap density at grain boundaries of TFTs remain as the major obstacles. Recently, the ultra-thin poly-Si TFT has been explored to significantly sharpen subthreshold slope and tighten V_{th} distribution by reducing the total number of traps [70].

Another important structural variation investigated has been the 1D ultra-narrow sensing channel, where sub-10 nm channel width have been fabricated in silicon-on-insulator (SOI) structures and by nano-scale carbon nanotubes (CNT) [71, 72]. One major difference between 1D and 2D channels with discrete charge storage is that the 1D current is mainly controlled by the maximum barrier in the channel, and can be modulated by a single charge-storage node near the 1D channel, known as the "bottleneck" effect [71]. In the 2D channel case, the least resistive path controls the current in the percolation process. The bottleneck effect enhances the memory window and the single-electron sensibility of the device. Another important advantage of 1D channel is its larger CR when combined with a 2D control gate. This is essentially due to the fringing fields and highly 3D electrostatics evident in the potential contour of a back-gated CNT FET in Fig. 7(a) [73]. The sharp potential gradient around the CNT indicates sharp potential drop between the CNT and the NC while the much gentle potential gradient between the gate and the NC. The potential cutline in Fig. 7(b) further reveals that a very high CR can be realized even with a thickness ratio between the control oxide and the tunnel oxide over 7. The superior electrostatic coupling together with the bottleneck effect allows very efficient P/E [72-73]. Therefore, the narrow-channel configuration deserves serious consideration for Flash memory operated at ultra-low P/E voltage.

11

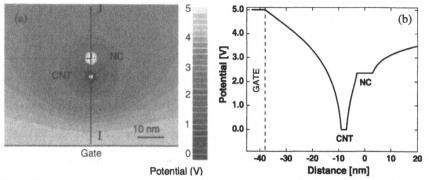

Figure 7 (a) Potential contour plot for a bottom-gate CNT FET with gate bias at 5V and CNT grounded, and (b) potential profile along the I-J cutline [73].

CONCLUSIONS

We have comprehensively reviewed the advantages and disadvantages in various material choices throughout the entire Flash device structure. A brief summary is provided in Table I. Despite tremendous technological challenges ahead for future FG nonvolatile memories, many viable solutions are also in sight. The heterogonous integration of new materials, such as metal NCs, high-κ trap layers, high-κ tunnel and control oxides, and metal control gates, enables new design space to increase the bit density and to optimize the memory performance at low P/E voltage. Moreover, the cost-effective 3D stacking and the innovative narrow-width channel provide new opportunities to go beyond the limitation imposed by the current planer memory array.

ACKNOWLEDGMENTS

This work is supported by National Science Foundation (NSF) through the Center of Nanoscale Systems (CNS) at Cornell University.

Table 1. Material selection in flash memory

Structure	Material	Main Advantages	Main Disadvantages	Note
Control gate	Poly-Si	- Conventional process	- Erase saturation	
	Metal	- Less erase saturation		TaN [62]
Control oxide	CVD SiO$_2$	- Conventional process	- EOT ↑ & coupling ratio ↓ - Channel control of NC ↓ - Coulomb energy of NC ↑ - P/E saturation	
	High-κ	- EOT ↓ & coupling ratio ↑ - Channel control of NC ↑ - Coulomb energy of NC ↓ - Less P/E saturation		Al$_2$O$_3$ [61-62], HfO$_2$ [35], HfAlO [23],
Charge storage	Poly-Si film	- Conventional process	- Floating gate crosstalk ↑ - SILC immunity ↓ - ψ drop	
	Dielectric traps	- Compatible process for nitride - Number density ↑	- EOT ↑ - Capture X-section ↓ - ψ drop with high-κ barriers - Compatibility of high-κ	Nitride [15-19], Al$_2$O$_3$ [20], HfAlO [21], HfSiO [22], AlN [23], HfO$_2$ [24], Ta$_2$O$_5$ [25]
	Semi. NC	- Compatible process	- ψ drop with high-κ barriers - Field diverge with high-κ - Small memory window - Number fluctuation	Si [26-29], Ge [30], C$_{60}$ [31]
	Metal NC	- Tunable workfunction - No ψ drop - Field converge with any dielectric	- Compatibility - Number fluctuation	Au, Pt, Ag [32-34], Ni [35], NiSi$_2$ [36], TiSi$_2$ [37], W, Co 40], C-tube [38]
Tunnel oxide	Thermal SiO$_2$	- Good endurance - Low D_{it}	- Nonscalable with retention time requirements	
	High-κ	- Improve t_R / t_{PE} ratio	- Endurance ↓	Nitride [17, 29], Al$_2$O$_3$, HfO$_2$ [35], HfAlO [30]
Channel	Bulk silicon	- Conventional process	- Short-channel effect	
	SOI/Fin	- Short-channel effect ↓ - V_{th} variation ↓ - Floating at retention	- Manufacturing cost	[65-67]
	TFT	- Stackable - Floating at retention	- V_{th} variation and poor S.S. in low-T poly and a-Si TFT	[68-70]
	Nanowire/ nanotube	- Coupling ratio ↑ - Memory window ↑ - Floating at retention	- Device yield ↓	Si NW [71], CNT [72]

REFERENCES

1. K. Lim, and J. Choi, in *Proc. Non-Volatile Semiconductor Memory Workshop*, 2006, p. 9.
2. K. Kinam, J.-H. Choi, J. Choi, and H.-S. Jeong, in *Proc. VLSI Technology (VLSI-TSA-Tech)*, 2005, p. 88.
3. R. Bez, and P. Cappelletti, in *Proc. VLSI Technology (VLSI-TSA-Tech)*, 2005, p. 84.
4. G. Atwood, *IEEE Trans. Device and Materials Reliability*, vol. 4, pp. 301, 2004.
5. International Technology Roadmap for Semiconductors, 2005 edition, [Online]. Available: http://public.itrs.net/
6. H. Pon, in *Proc. Int. Solid-State and Integrated Circuit Technology*, 2006, p. 697.

7. J.-H. Kim *et al.*, in *IEDM Tech. Dig.*, 2006, p. 45.
8. S. Tehrani, in *IEDM Tech. Dig.*, 2006, p. 585.
9. J. H. Oh *et al.*, in *IEDM Tech. Dig.*, 2006, p. 49.
10. Y. Shin, in *Proc. VLSI Circ. Dig.*, 2005, p. 156.
11. L. Geppert, *IEEE Spectrum*, vol. 40, p. 48, 2003.
12. N. Flaherty, *IEE Review*, vol. 49, pp. 50, 2003.
13. M. Noguchi *et al.*, in *IEDM Tech. Dig.*, 2007, p. 445.
14. B. DeSalvo *et al.*, *IEEE Trans. Device Mater. Rel.*, vol. 4, p. 377, 2004.
15. H. A. R. Wegener *et al.*, in *IEDM Tech. Dig.*, 1967, p. 70.
16. C. T. Swift *et al.*, in *IEDM Tech. Dig.*, 2002, p. 927.
17. H.-T. Lue *et al.*, in *IEDM Tech. Dig.*, 2005, p. 547.
18. Y. Park *et al.*, in *IEDM Tech. Dig.*, 2006, p. 29.
19. R. Ohba, Y. Mitani, N. Sugiyama, and S. Fujita, in *IEDM Tech. Dig.*, 2007, p. 75.
20. T. Sugizaki, *et al.*, in *Symp. VLSI Tech.*, 2003, p. 27.
21. Y. Tan, W. Chim, W. Choi, M. Joo, T. Ng, and B. Cho, in *IEDM Tech. Dig.*, 2004, p. 889.
22. Y.-H. Lin, C.-H Chien, C.-T. Lin, C.-W. Chen, C.-Y. Chang, and T.-F. Lei, in *IEDM Tech. Dig.*, 2004, p. 1080.
23. C. H. Lai, *et al.*, in *Symp. VLSI Tech.*, 2005, p. 210.
24. Y. Q. Wang *et al.*, in *IEDM Tech. Dig.*, 2006, p. 971.
25. X. Wang, J. Liu, W. Bai, and D.-L. Kwong, *IEEE Trans. Electron Devices*, vol. 51, p. 597, 2005.
26. S. Tiwari, F. Rana, K. Chan, H. Hanafi, W. Chan, and D. Buchanan, in *IEDM Tech. Dig.*, 1995, p. 521.
27. R. Muralidhar *et al.*, in *IEDM Tech. Dig.*, 2003, p. 601.
28. B. De Salvo *et al.*, in *IEDM Tech. Dig.*, 2003, p. 597.
29. S. Baik, S. Choi, U. I. Chung, and J. T. Moon, in *IEDM Tech. Dig.*, 2003, p. 545.
30. J. H. Chen *et al.*, *IEEE Trans. Electron Devices*, vol. 51, p. 1840, 2004.
31. T. H. Hou, U. Ganguly, and E. C. Kan, *Appl. Phys. Lett.*, vol. 89, 253113, 2006.
32. Z. Liu, C. Lee, V. Narayanan, G. Pei, and E. C. Kan, *IEEE Trans. Electron Devices*, vol. 49, p. 1606, 2002.
33. C. Lee, A. Gorur-Seetharam, and Edwin C. Kan, in *IEDM Tech. Dig.*, 2003, p. 557.
34. C. Lee, J. Meteer, V. Narayanan, and E. C. Kan, *J. Electronic Materials*, vol. 34, p. 1, 2005.
35. J. J. Lee, and D.-L. Kwong, *IEEE Trans. Electron Devices*, vol. 52, p. 507, 2005.
36. P. H. Yeh *et al.*, *J. Vac. Sci. Technol. A*, vol. 23, p. 851, 2005.
37. Y. Zhu, D. Zhao, R. Li, and J. Liu, *Appl. Phys. Lett.*, 88, 103507, 2006.
38. X. B. Lu and J. Y. Dai, *Appl. Phys. Lett.*, 88, 113104, 2006.
39. R. Ohba, N. Sugiyama, K. Uchida, J. Koga, and A. Toriumi, *IEEE Trans. Electron Devices*, vol. 49, p. 1392, 2002.
40. M. Takata *et al.*, in *IEDM Tech. Dig.*, 2003, p. 553.
41. S. Tang, C. Mao, Y. Liu, D. Q. Kelly, and S. K. Banerjee, in *IEDM Tech. Dig.*, 2005, p. 181.
42. K. W. Guarini, C. T. Black, Y. Zhang, I. V. Babich, E. M. Sikorski, and L. M. Gignac, in *IEDM Tech. Dig.*, 2003, p. 541.
43. C.N.R. Rao, G. U. Kulkarni, P. J. Thomas, and P. P. Edwards, *Chem. Eur. J.*, vol. 8, p. 29, 2002.
44. M. She, and T.-J. King, *IEEE Trans. Electron Devices*, vol. 50, p. 1934, 2003.
45. Y. Liu, S. Tang, and S. K. Banerjee, *Appl. Phys. Lett.*, 88, 213504, 2006.

46. Y. Yang, and M. H. White, *Solid-state Electron.*, vol. 44, p. 949, 2000.
47. C. Lee, U. Ganguly, V. Narayanan, T.-H. Hou, and E. C. Kan, *IEEE Electron Device Lett.*, vol. 26, p. 879, 2005.
48. T.-H Hou, C. Lee, V. Narayanan, U. Ganguly, and E. C. Kan, *IEEE Trans. Electron Devices*, vol. 53, p. 3095, 2006.
49. T.-H Hou, C. Lee, V. Narayanan, U. Ganguly, and E. C. Kan, *IEEE Trans. Electron Devices*, vol. 53, p. 3103, 2006.
50. T. H. Hou, C. Lee, and E. C. Kan, in *Device Research Conf. Dig.* 2007, p.221.
51. L. Perniola *et al.*, *IEEE Trans. Nanotechnol.*, vol. 2, p. 277, 2003.
52. R. Gusmeroli, C. M. Compagnonia and A. S. Spinellia, *Microelectronic Eng.*, vol. 84, p. 2869, 2007.
53. S. Huang, K. Arai, K. Usami, and S. Oda, *IEEE Trans. Nanotech.*, vol. 3, p. 210, 2005.
54. C. Lee, T-H Hou, and E. Kan, *IEEE Trans. Electron Devices*, vol. 52, p. 2697, 2005.
55. N. Yang, W. K. Henson, J. R. Hauser, and J. J. Wortman, *IEEE Trans. Electron Devices*, vol. 46, p. 1464, 1999.
56. K. K. Likharev, *Appl. Phys. Lett.* vol. 73, p. 2137, 1998.
57. P. Blomme, B. Govoreanu, M. Rosmeulen, J. Van Houdt, and K. DeMeyer, *IEEE Electron Device Lett.*, vol. 24, p. 99, 2003.
58. C. M. Compagnoni, D. Ielmini, A. S. Spinelli, and A. L. Lacaita, *IEEE Trans. Electron Devices*, vol. 52, p. 2473, 2005.
59. M. V. Fischetti, D. A. Neumayer, and E. A. Cartier, *J. Appl. Phys.*, vol. 90, p. 4587, 2001.
60. K. Mistry *et. al*, in *IEDM Tech. Dig.*, 2007, p. 247.
61. W.-H. Lee, J. T. Clemens, R. C. Keller, and L. Manchanda, in *Symp. VLSI Tech. Dig.*, 1997, p. 117.
62. C. H. Lee, K. I. Choi, M. K. Cho, Y. H. Song, K. C. Park, and K. Kim, in *IEDM Tech. Dig., 2003*, p. 613.
63. J. Robertson, *J. Vac. Sci. Technol. B*, vol. 18, p. 1785, 2000.
64. T. H. Hou, U. Ganguly, and E. C. Kan, *IEEE Electron Device Letters*, vol. 28, p.103, 2007.
65. D. Burnett, D. Shum, and K. Baker, in *IEDM Tech. Dig.*, 1998, p. 983.
66. P. Xuan, M. She, B. Harteneck, A. Liddle, J. Bokor, and T.-J. King, in *IEDM Tech. Dig.*, 2003, p. 609.
67. H. Silva, M.K. Kim, A. Kumar, U. Avci, S. Tiwari, in *IEDM Tech. Dig.*, 2003, p. 271.
68. S.-M. Jung *et al.*, in *IEDM Tech. Dig.*, 2006, p. 37.
69. E.-K. Lai *et al.*, in *IEDM Tech. Dig.*, 2006, p. 41.
70. Y. Fukuzumi *et al.*, in *IEDM Tech. Dig.*, 2007, p. 449.
71. M. Saitoh, E. Nagata and T. Hiramoto, *Appl. Phys. Lett.*, vol. 82, p. 1787, 2003.
72. U. Ganguly, E. C. Kan and Y. Zhang, *Appl. Phys. Lett.*, vol. 87, 43108, 2005.
73. U. Ganguly, C. Lee, T. H. Hou, and E. C. Kan, *IEEE Trans. Nanotech.*, vol. 6, p. 22, 2007.

Mater. Res. Soc. Symp. Proc. Vol. 1071 © 2008 Materials Research Society   1071-F02-02

Realization of Hybrid Silicon core/silicon Nitride Shell Nanodots by LPCVD for NVM Application

Jean Colonna[1], Gabriel Molas[1], Marc W Gely[1], Marc Bocquet[1], Eric Jalaguier[1], Barbara De Slavo[1], Helen Grampeix[1], Pierre Brianceau[1], Karim Yckache[1], Anne-Marie Papon[1], Geoffroy Auvert[1], Corrado Bongiorno[2], and Salvatore Lombardo[2]

[1]CEA-Léti MINATEC, 17, Avenue des Martyrs, grenoble, 38054, France
[2]IMM CNR, Stradale Primosole,50, Catania, 95121, Italy

ABSTRACT

We present the realization of hybrid silicon core/silicon nitride shell nanodots by Low Pressure Chemical Vapor Deposition (LPCVD) and their application as floating gate in Non Volatile Memory (NVM) devices. The LPCVD process includes three steps: nucleation using SiH_4, selective growth of the silicon nuclei using SiH_2Cl_2 and finally selective growth of silicon nitride using a mixture of SiH_2Cl_2 and NH_3 around the silicon dot. The two first steps have already been described in literature. We will therefore focus on the selective growth of a nitride layer on silicon dots. Morphological characterization using Scanning Electron Microscopy (SEM) allows control over dots size – 5 to 10nm – and density – up to 1E12/cm². High Resolution Transmission Electron Microscopy (HRTEM) shows a crystalline silicon core and an outer shell of amorphous silicon nitride. Energy Filtered TEM pictures confirm that the nitride layer is deposited only around the silicon dots and not on the oxide. Oxidation resistance of the silicon nitride shell is also investigated. A 2nm thick silicon nitride layer is an efficient barrier to an oxidation at 800°C in dry oxygen for 5 minutes. We thus have a very thin high quality stoichiometric nitride layer. Such a high quality nitride film can only be achieved using in-situ deposition i.e. on an oxide-free silicon surface. Finally, hybrid Si/SiN nanodots are integrated in a single memory cell with high-K interpoly dielectric. Electrical results show large threshold voltage shift of 6V. The use of silicon nitride shells on the silicon dots has therefore two main advantages: it provides both oxidation resistance and charge storage enhancement.

INTRODUCTION

The market of flash NAND memories is considerably increasing nowadays due to the large success of mobile electronic devices. Discrete trap memories such as SONOS and Silicon nanocrystals memories are among the best candidates for future technologies. However, one of the main issues for silicon nanocrystals memories is their relatively small programming window. Besides, the integration of such nanocrystals raises the problem of their potential oxidation during the following steps of the process flow. As a possible solution to both issues, we propose the addition of a silicon nitride shell around the silicon nanocrystals. This paper describes the realization of such hybrid nanodots in an industrial Low Pressure Chemical Vapour Deposition furnace.

EXPERIMENT AND DISCUSSION

Growth of hybrid Silicon/Silicon nitride nanodots by LPCVD

The hybrid nanodots are grown on the tunnel dielectric in case of Non Volatile Memories (NVM). Most commonly, it will be silicon thermal oxide. In order to improve the nanodots density, a chemical clean (RCA or HF-RCA) is recommended. It will increase the number of silanol groups (-OH) on the surface of the SiO2. Those silanol groups are used as nucleation sites [1].

The Low Pressure Chemical Vapor Deposition (LPCVD) process can be described in three steps. It includes the growth of silicon nanocrystals (step 1 and 2) and their in-situ passivation by deposition of a silicon nitride layer around the nanocrystals (step 3) in the same tool as shown in fig. 1. The benefit of step 1 and 2 is to increase density and reduce nanocrystals size dispersion compared to a standard one step silane growth [2]. Concerning step 3, the requirements regarding the nitride layer are: a thin layer (few nanometers) that shows good resistance to oxidation and selectivity, which means nitride deposition only around the silicon nanocrystals.

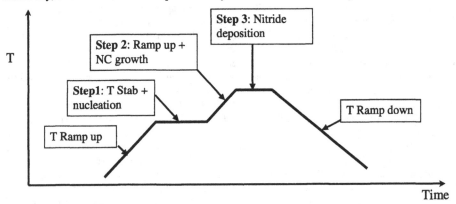

Figure 1: schematic of the 3-step LPCVD recipe. Step 1: nucleation SiH₄, step 2: Silicon nucleus growth SiH2Cl2 (DCS), step 3: selective nitride deposition.

Selectivity:
This phenomenon is also known in literature as nucleation delay or growth retardation [3]. The delay is maximum on thermal oxide - approximately 8 minutes for standard LPCVD nitride and is shorter on native oxide - approximately 5 minutes. It is almost reduced to zero on an oxide-free silicon surface (after an "HF-last" wet cleaning for example). The surface of a thermal oxide has siloxane bonds -Si-O-Si- which represents the worse case for nucleation. Native oxide surfaces have a majority of silanol bonds Si-OH which are more advantageous for nucleation. The best case is an oxide-free silicon surface with Si-H bonds.

In the case of silicon nanocrystals grown on thermal oxide, in-situ deposition - i.e. silicon dots with an oxide-free silicon surface - allows the nitride deposition to be selective for deposition times shorter than 8 minutes.

Oxidation resistance of thin silicon nitride layers:
Thick silicon nitride layers are known to be efficient oxidation barriers. However this is not the case below a certain limit thickness [4]. This is due to a transition layer at the initial stage of deposition when nitride is deposited on oxide. This transition layer is non-stoechimetric (Si-rich)

and shows a weak oxidation resistance. To reduce this transition layer, silicon nitride can be deposited on oxide-free silicon surfaces. This is the case on silicon nanocrystals. The silicon nitride shell grown on silicon dots with an oxide-free silicon surface is therefore stoechiometric and shows good oxidation resistance.

Morphological characterization

In order to verify that the nitride shell is an efficient oxidation barrier, we deposited the hybrid nanodots on a 10nm thick nitride layer. A Scanning Electron Microscopy (SEM) picture of the hybrid nanodots is shown fig 2-a. We then oxidized the nanodots in a standard oxidation furnace at 800°C in dry oxygen for 5 minutes. The oxide thickness measured on Silicon monitor wafer is 3.2nm. After the oxidation, the nanodots were etched in HF (HF: 0,2%, HCl: 1% etch rate: 11Å/min on thermal oxide) for 5 minutes. A SEM picture of the hybrid nanodots after oxidation and HF etch is shown fig 2-b. From this picture, it can be established that the nanodots have not been oxidized. As a reference, the same experiment was performed on large silicon dots. Fig 2-c and fig 2-d show these silicon dots before and after oxidation and HF etch. It confirms that, with our experimental protocol, bare silicon dots are oxidized and the HF etch removes them.

The results clearly show that the nitride shell is an efficient barrier to oxidation at 800°C in dry oxygen for 5 minutes.

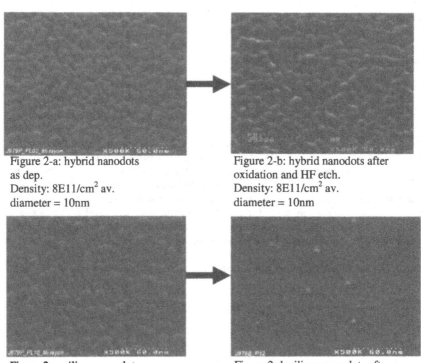

Figure 2-a: hybrid nanodots
as dep.
Density: $8E11/cm^2$ av.
diameter = 10nm

Figure 2-b: hybrid nanodots after
oxidation and HF etch.
Density: $8E11/cm^2$ av.
diameter = 10nm

Figure 2-c: silicon nanodots
as dep.
Density: $4E11/cm2$ av.
diameter = 12nm

Figure 2-d: silicon nanodots after
oxidation and HF etch

Energy Filtered Transmission Electron Microscopy (EFTEM) pictures are shown in fig 3. The hybrid nanodots were deposited on a 4nm thermal oxide on this sample. Fig 3-a is an EFTEM picture at 16eV, which is the energy of the Si plasmon excitation. This picture therefore shows in bright the silicon core of the hybrid nanodots. Fig 3-b shows a picture at 25eV, which reveals both SiO_2 and Si_3N_4. The silicon core therefore appears dark, the nitride shell is bright and the oxide between dots appears slightly darker than the nitride. This picture also allows us to estimate the nitride shell thickness: it is approximately 2nm thick. Fig 3-c shows a nitrogen map. The intensity between the dots goes down to zero, clearly showing that there is no nitride on the tunnel oxide between dots.

 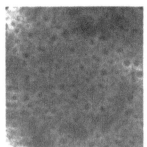 

Figure 3-a: EFTEM picture at 16eV showing Silicon in bright	Figure 3-b: EFTEM picture at 25eV. Silicon core appears as dark full circles, nitride shells around Si core are bright and SiO_2 between dots is grey	Figure 3-c: Nitrogen map. No nitrogen detected on the oxide between dots

Fig 4-a shows a SEM picture of an optimized process with the appropriate nanodot size and density for NVM applications. Silicon core has an average diameter of 7nm, the nitride shell is 1.5nm thick and the density is 9E11dots/ cm^2.

 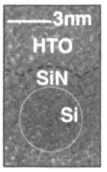

Figure 4-a: optimized process Density: 9E11/cm2 av. diameter = 10nm

Figure 4-b: TEM picture of optimized process with HTO capping. The Si core is clearly crystalline.

Fig4-b shows a High Resolution TEM picture of a hybrid nanodot with High Temperature Oxide (HTO) capping. The silicon core is clearly crystalline and the outer nitride shell amorphous. Morphological characterizations do confirm our theoretical considerations on selectivity – see fig. 3-c and oxidation resistance – experiment described in fig. 2.

Electrical characterization

The hybrid Si/SiN nanodots corresponding to fig 4-a were integrated as a floating gate in a memory cell [5]. The schematic of the memory device is presented in fig 5-a. It includes an interpoly dielectric composed of the following stack: HTO 4nm thick / HfAlO high-k layer 8nm thick / HTO 4nm thick. The write/erase characteristics of such a device are presented in fig 5-b. The write curves in red were acquired with voltage ranging from 14 to 18V and the erase curves in black with voltage from -12 to -15V. If we consider a writing time of 1ms, a writing voltage and respectively erasing Vw=18V and Ve=-15V, the threshold voltage shift is 6V, which is clearly noticeable.

Another electrical characteristic is the feasibility of 2-bits cell operation in hot carrier programming mode. A 25% shift on the threshold voltage between the two bits was demonstrated (electrical graph not shown here). It means that when operating hot carrier injection from the drain, the charge is injected in the dots near the drain region and stays localized there. This clearly demonstrates that the hybrid nanodots shown in fig 4-a act as discrete traps.

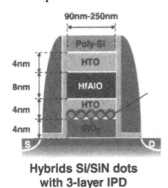

Hybrids Si/SiN dots with 3-layer IPD

Figure 5-a: schematic of memory device

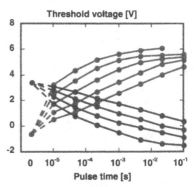

Figure 5-a: write/erase characteristic of hybrid Si/SiN memory

ACKNOWLEDGMENTS

Part of this work was supported by the MEDEA+ NEMESYS project.

REFERENCES

1. F. Mazen et al. Journ. of Electr. Chem. Society, 150 (3) 203, 2003
2. F. Mazen et al. Applied Surface Science 214, p359-363, 2003
3. F. Martin et al. Semicond. Sci. Technol. 6, 1100, 1991
4. M. Yoshimaru et al. IEEE Transactions on Electron Devices, vol 41, n° 10, 1994
5. G Molas et al. To be presented at IEDM 2007, Washington DC.

Mater. Res. Soc. Symp. Proc. Vol. 1071 © 2008 Materials Research Society 1071-F02-05

Impact of the High-Temperature Process Steps on the HfAlO Interpoly Dielectric Stacks for Nonvolatile Memory Applications

Daniel Ruiz Aguado[1,2], Bogdan Govoreanu[1], Paola Favia[1], Kristin De Meyer[1,2], and Jan Van Houdt[1]

[1]IMEC, Kapeldreef 75, Leuven, 3001, Belgium
[2]K. U. Leuven, Kasteelpark Arenberg 10, Leuven, 3001, Belgium

ABSTRACT

This work reports on the performance of different Hafmiun aluminate ($HfAlO_x$)-based interpoly dielectrics (IPD) for future sub-45nm nonvolatile memory (NVM) technologies. The impact of the thermal budget during the fabrication process is studied. The good retention and large operating window shown by this material, can be compromised by a high temperature activation anneal (AA) after the gate deposition. The AA step may induce phase segregation of the $HfAlO_x$ and outdiffusion of the Hf (Al) towards the floating gate/IPD and IPD/gate interfaces and subsequent formation of Hf (Al) silicates. These findings are supported by the low field leakage analysis, which shows large device to device dispersions. However, the effect of the spike anneal can be minimized if the $HfAlO_x$ layer is crystallized prior to the AA. Devices with polysilicon or TiN gate are compared in terms of memory performance and reliability.

INTRODUCTION

Scaling Flash memory technology to sub-45nm nodes requires cell planarization; hence a considerable reduction of the electrical thickness of the IPD is required in order to compensate for the loss of the sidewall coupling capacitance. This is achievable by combining high-k IPD's with high-work-function metal gates [1-3].

When used as IPD, $HfAlO_x$ shows a large program/erase (P/E) window, while keeping the operating voltages at acceptable levels. Moreover, a good room temperature (RT) retention was observed for very thin layers, of only 12 nm physical thickness, with 0.5V window closure after more than 10^6s. This work reports on the impact of the thermal budget during the fabrication process on the performance of different $HfAlO_x$-based IPD's. The impact on performance of the gate material is also discussed.

The paper is organized as follows: in the second section, a description of the test structures, review of the process flow and summary of the studied samples is given. The third section discusses the electrical results, focusing on retention ability of the IPD's stacks, as well as on their physical characterization.

TEST STRUCTURES

The test structures are fabricated in a process that allows formation of both floating gate capacitors and IPD capacitors on the same wafer.

The floating gate capacitors have 8.5nm thermal SiO_2 tunnel oxide (TO) grown on p-type Si substrate and a floating gate (FG) of n^+ poly-Si formed by in-situ P-doping, on top of the oxide. The IPD stack is formed on top of the floating gate (FG), following a chemical-mechanical polishing (CMP) of the poly-Si film. This stack consists of a thin (~1nm) SiO_2 interface and a 12nm layer of $HfAlO_x$ (high-k layer) formed by atomic layer deposition (ALCVD) with an Al_2O_3:HfO_2 deposition cycle ratio of 1:1. Before the metal gate deposition, a degas process is performed at 500°C during 3 minutes, in order to desorb water residues possibly present in the high-k layer. Different post-deposition-anneal (PDA) conditions were used, with temperatures ranging between 700°C and 1100°C, for 60s, or as a 10s spike anneal. The control gate (CG) is formed by depositing either n^+ poly-Si, TiN, deposited from ionized metal plasma (IMP) [4], or a metal-inserted poly-Si (MIPS), consisting of 10nm of IMP TiN and 100nm n^+ poly-Si on top. All splits, except N7, received a spike AA at 1030°C during 10s after the gate deposition.

The IPD capacitors have the previously mentioned IPD stacks as dielectric between the Si substrate and the gate. The IPD stacks discussed in this work are summarized in Table I.

Table I. Summary of the IPD stacks and their corresponding high-temperature process steps.

Sample	IPD stack	PDA conditions	Gate	Activation anneal
S1	1nm SiO_2/12nm $HfAlO_x$	800°C, 1 min	n^+-Poly	Yes
S2	1nm SiO_2/12nm $HfAlO_x$	700°C, 1 min	MIPS	Yes
S3	1nm SiO_2/12nm $HfAlO_x$	800°C, 1 min	MIPS	Yes
S4	1nm SiO_2/12nm $HfAlO_x$	900°C, 1 min	MIPS	Yes
S5	1nm SiO_2/12nm $HfAlO_x$	1000°C, 10s (spike)	MIPS	Yes
S6	1nm SiO_2/12nm $HfAlO_x$	1100°C, 10s (spike)	MIPS	Yes
N7	1nm SiO_2/12nm $HfAlO_x$	800°C, 1 min	TiN	No

RESULTS AND DISCUSSION

The next two sections discuss the performance of the splits in terms of P/E window and retention, respectively. Large-area floating-gate capacitors were programmed (erased) by applying a positive (negative) pulse to the control gate. The flatband-voltage (V_{FB}) shift in the high-frequency capacitance-voltage (HFCV) curves was used to monitor the P/E window.

P/E operating window

A large P/E window, up to ~9V, is achievable in the split without AA, N7 (Figure 1). The dashed lines of Figure 1 show the theoretical trends corresponding to P/E characteristics where the net FG charge is modified due to injection through the TO only. The V_{FB}-shift with applied erase pulse voltage follows the theoretical trends closely, suggesting good immunity to erase saturation down to -6V of V_{FB} shift. This is a result of the midgap to p-type character of the TiN gate, determining a large barrier height at the CG/high-k interface. In contrast, strong saturation is observed during programming due to parasitic injection of electrons through the IPD.

A remarkable difference is observed between the split N7 and the rest of the splits. The high-temperature AA reduces the operating window by ~1.5V in the programmed state and ~1 volt in erased state. Furthermore, earlier erase saturation is reached with the split with poly-Si gate. This difference is due to the different work function of the poly-Si with respect to the MIPS.

24

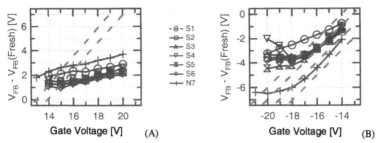

Figure 1. V_{FB}-shift of the splits from Table I vs. program (A) and erase (B) voltages applied to the CG, for a 10ms pulse time. The dashed lines indicate the ideal slope of the curves.

Retention at different temperatures

To monitor the room temperature (RT) retention (Figure 2), the P/E voltages were selected to give a sufficiently large window without inducing degradation in the structure.

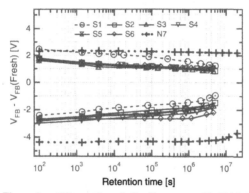

Figure 2. RT retention of all splits presented in Table I.

The split without AA, N7, shows significant differences with respect to its counterparts with AA. No initial window closure is observed in this split, which suggests less trapping in the IPD. The total window closure after 10^6s, attributed to FG charge loss, is less than 0.5V. This charge loss is lower compared to the other splits, even if the initial P/E V_{FB} values are larger.

All the other splits, with AA, show some initial window closure, of more than 1V, after the first 10^4s. This initial window closure is attributed to charge detrapping from the IPD.

In the programmed state all the splits with AA and MIPS gate show similar results. Only the split with poly-Si gate (S1) shows larger window closure than the splits with MIPS, although this can be due to the higher initial programmed-V_{FB} value.

In the erased state, all the S splits, except the split with a PDA at 1100°C (S6), show similar retention results, with ~1.5V window closure after 10^6 seconds. S6 shows a much flatter curve with ~0.2V window closure after 10^6s.

For a better understanding of the observed retention behavior, retention tests at RT, 150°C and 250°C are shown in Figure 3. The splits with different gate, in Figure 3 (A), show no important differences between them, probably because the impact of the gate is overridden by the AA. Figure 3 (B) shows that the split S6, with 1000°C PDA, has almost total window closure, even after only 30min, at 250°C and 3V of closure after 2 hours at 150°C. Similar results were obtained for the splits with lower PDA temperatures (Figure 3 (A) shows the retention of the split with 800°C PDA). On the other hand, the split with a PDA at 1100°C presents much less charge loss acceleration with the temperature. Figure 3 (C) shows that the previous results obtained from the splits with AA are always worse than the ones obtained with the split without AA. The split without AA has lower window closure at 150°C and 250°C in both programmed and erased state, which indicates a strong impact of the activation anneal on the final HfAlO$_x$-based IPD stack.

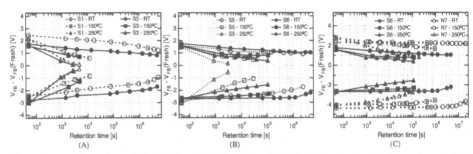

Figure 3. Retention at RT, 150°C and 250°C of the splits with different gates (A): S1 - poly gate, S3 - MIPS gate, with different PDA conditions (B): S5 = PDA 1000°C; S6 = PDA 1100°C and with or without AA (C): S6 = with AA, N7 = without AA.

IPD stack analysis

Transmission electron microscopy (TEM) pictures (not shown) indicate that the HfAlO$_x$ layer remains amorphous after the PDA at 800°C. However, crystalline features are found in all the splits with a PDA at 800°C, suggesting that the AA at 1030°C crystallizes the HfAlO$_x$ layer. Thus, all the splits with AA have a polycrystalline HfAlO$_x$ IPD.

A scanning transmission electron microscopy (STEM) picture of the split without AA is shown in Figure 4 (A). No interfacial layer (IL) appears between the HfAlO$_x$ and the TiN of the gate. Moreover, the bottom SiO$_2$ interface layer is quite uniform and its thickness is about 1nm, as expected. STEM pictures of the splits with AA, with 800°C PDA, S3, and with 1100°C PDA, S6, are presented in Figure 4 (B) and (C), respectively.

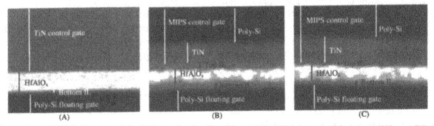

Figure 4. STEM pictures of the IPD stack of split without AA, N7 (a), and with AA at different PDA temperatures: at 800°C, S3 (B), and at 1100°C, S6 (C).

It has been shown in Figure 2 and 3 that the splits with AA have worse retention results than the split without AA. Based on the STEM pictures' contrast and on the atomic composition analysis performed on the stacks with AA, two main phenomena could explain those electrical results. First, phase segregation of the $HfAlO_x$ into Hf-rich and Al-rich phases, likely occurs. Second, the Hf and Al seem to mix with the bottom SiO_2 interface layer and with the top TiN gate. Due to this mixing, silicate compounds are most likely formed at the SiO_2/high-k interface and HfTiON compounds are also formed at the top high-k/gate interface.

Furthermore, the main electrical differences between the splits with AA, S3 and S6, are mainly observed in the erased state. Consequently, it is expected that the main structural differences will be found in the interface between the IPD and the gate. Figure 4 (B) and (C) show that a thick top interface between the $HfAlO_x$ and the top TiN is formed by the AA independently from the PDA temperature. In the split S3, this additional parasitic top interface is formed by two layers: an Al-rich layer in contact with the $HfAlO_x$ and a Hf-rich layer on top, in contact with the TiN of the gate, probably forming HfTiON compounds. The thickness of the Al-rich IL is ~1.5nm and the Hf-rich layer is ~1nm. However, the split with higher PDA temperature presents only the Al-rich layer as parasitic interface between the $HfAlO_x$ and the TiN. The thickness of this Al-rich layer is ~1.5nm too.

A $HfAlO_x$ film may need a crystallization temperature above 1000°C if the content of Al is superior to 45% [5], which is the case for the layers used in this work. Moreover, all the splits with AA and PDA temperatures ranging 700°C to 1000°C show similar electrical results. This suggests that there is no significant structural difference among them. Since it has also been observed that a PDA at 800°C during 1 minute does not crystallize the $HfAlO_x$ layer, then, most likely the 1000°C PDA during 10 seconds does not crystallize the layer either. 1000°C is then considered as the temperature limit above which crystallization occurs, since the 1030°C of the AA crystallized the $HfAlO_x$ layer. Consequently, the PDA at 1100°C crystallized the $HfAlO_x$ layer before the gate formation and the AA step. This crystallization before the high-temperature AA reduced its negative impact on the properties of the structure.

J-V characteristics (not shown) were measured on IPD capacitors with the same dielectric structure as the IPD of the splits S3 and S6. The J-V curves indicate a higher leakage current occurs at all voltages in the split S3, especially for the negative polarity, where the Hf-rich layer impact is expected to be stronger. The formed top interface layers and the more prominent phase separation observed in the splits that were non-crystalline before the AA (Figure 4 (B) and (C)) can explain the electrical differences observed between the splits S3 and S6.

In order to further probe the previous observations, the low-field leakage current was extracted from the variations of erased V_{FB} observed in the retention measurements and then fitted with simulations [6].

The layers defined for the extraction of the leakage and the simulations are: 1nm of SiO_2 bottom interface and 12nm of $HfAlO_x$. The parasitic $HfAlO_x$/gate IL found in the splits with AA are still considered as $HfAlO_x$. Due to their small thickness, possible small differences in the band offset or dielectric constant may be neglected. However, additional traps are considered in that interface to follow the observations from the STEM pictures of Figure 4.

The leakage current density extracted from the split S6 can be well fitted considering a conduction band (CB) offset (relative to Si) of 2eV, which is typical for the $HfAlO_x$ [7, 8], a uniform distribution of traps in an energy band around 0.8 ± 0.1eV below the high-k conduction band, with a volume density of $2.5 \cdot 10^{19}/cm^3$ and a higher density of traps of $2.5 \cdot 10^{20}/cm^3$, at a energy band around 1.0 ± 0.1eV in the top IL.

The leakage current density extracted from the split S3 can be well fitted considering a CB offset of 1.6eV and the same parameters for the trap density and energy levels as for S6. The phase segregation into Al and Hf-rich crystalline grains can explain the lower CB offset in this case. A Hf-rich grain will have a lower conduction band, presumably closer to the HfO_2 value, which will dominate the leakage due to the exponential dependence on the barrier height.

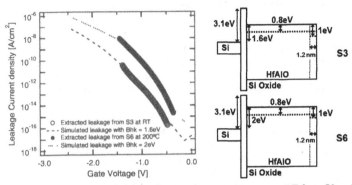

Figure 5. Leakage current extracted from the retention measurements at RT from S3 and at 200°C from S6, fitted with simulations using the indicated band diagrams, thicknesses and trap levels.

CONCLUSIONS

This work reports on the performance of different $HfAlO_x$-based interpoly dielectrics for future sub-45nm nonvolatile memory (NVM) technologies. The impact of the thermal budget during the fabrication process has been studied. The results indicate that the good retention and large operating window shown by this material can be compromised by a high-temperature spike activation anneal after the gate deposition. The activation anneal leads to phase segregation and intermixing of the $HfAlO_x$ with the bottom SiO_2 interface layer and creates parasitic interfacial layers in the top IPD/gate top interface. This additional interfacial layer presents a higher density of defects as shown by the I-V measurements. However, the effect of the activation anneal can be minimized if the $HfAlO_x$ layer is crystallized before the activation anneal. The impact of the gate material is also discussed. Structures with polysilicon gate show smaller P/E window due to the higher Fermi level of the n^+-type polysilicon. However, the retention is similar to the structures with TiN or MIPS gates, since the effect of the spike activation anneal largely determine the final quality of the structure.

REFERENCES

[1] B. Govoreanu, D. P. Brunco, J. Van Houdt, Solid-St. Electr., 49(11): 1841-1847, 2005.
[2] M. van Duuren, R. van Schaijk, M. Slotboom, P. Tello, P. Goarin, N. Akil, F. Neuilly, Z. Rittersma and A. Huerta, Proc. NVSM Workshop, pp. 48-49, 2006.
[3] D. Wellekens, P. Blomme, B. Govoreanu, J. De Vos, L. Haspeslagh, J. Van Houdt, D. P. Brunco and K. van der Zanden, Proc. ESSDERC, pp. 238-241, 2006.
[4] D. Dictus, D. Shamiryan, V. Paraschiv, and W. Boullart, S. De Gendt and S. Vanhaelemeersch, J. Vac. Sci. Technol. B 24(5): 1071-1023, 2006.
[5] W. J. Zhu, T. Tamagawa, M. Gibson, T. Furukawa, and T. P. Ma, IEEE El. Dev. Lett, 23(11): 649-651, 2002.
[6] B. Govoreanu, D. Wellekens, J. De Vos, L. Haspeslagh, J. Van Houdt, IEDM Tech. Dig, pp. 479-482, 2006.
[7] H. Y. Yu, N. Wu, M. F. Li, C. Zhu and B. J. Cho, Appl. Phys. Lett., 81(19): 3618-3620, 2002.
[8] V. V. Afanas'ev, A. Stesmans and W. Tsaib, Appl. Phys. Lett., 82(2): 245-247, 2003.

Mater. Res. Soc. Symp. Proc. Vol. 1071 © 2008 Materials Research Society   1071-F02-08

Effect of Top Dielectric Morphology and Gate Material on the Performance of Nitride-based FLASH Memory Cells

Antonio Cacciato, Laurent Breuil, Geert Van den bosch, Olivier Richard, Aude Rothschild, Arnaud Furnémont, Hugo Bender, Jorge A. Kittl, and Jan Van Houdt
IMEC, Kapeldreef 75, Leuven, B-3001, Belgium

ABSTRACT

In this paper we study the effect of the gate material and of the top dielectric morphological transformation associated with the high-k post deposition anneal on the erase and the retention behavior of nitride-based memory cells. In particular, we show that for Al_2O_3 and HfAlO a trade-off exists between erase and retention, higher PDA temperatures being beneficial for erase but detrimental for retention. We also discuss the effect of Fermi level pinning and poly-Si depletion on the erase behavior and compare the erase performances of several PVD (Physical Vapor Deposition)- and AVD (Atomic Vapor Deposition)-deposited metal gates.

INTRODUCTION

The nitride-based SONOS cell, for its excellent scalability and process simplicity, is the candidate to push the scaling roadmap for FLASH memories beyond the limit imposed on floating-gate memories by the electrostatic interference between adjacent cells [1]. The traditional SONOS cell consists of a nitride layer (the storage element) encapsulated by two SiO_2 layers which isolate the nitride layer from the Si substrate and the poly-Si gate (Poly-Si/SiO_2/Si_3N_4/SiO_2/c-Si). However, the thick tunnel oxide necessary to meet the retention requirements imposes a severe limit on the erase performance because of the erase saturation phenomenon [2]. One possibility to guarantee both the erase and the retention performance is the replacement of the top SiO_2 layer with materials of higher dielectric constant (high-k dielectric). The presence of a high-k dielectric reduces the electric field across the top dielectric, thus decreasing the unwanted parasitic electron injection from the gate during the erase operation [2]. This will allow the cell to erase deep so to meet a basic requirement for Gigabit multilevel NAND memories [7]. The introduction of high-k materials in the SONOS stack is unfortunately not straightforward. One problem is the Fermi-level pinning at the poly-Si/high-k interface [3]. Another problem is the morphological changes the high-k material undergoes during the device fabrication thermal budget. These changes can modify the k-value and affect the band offset between gate and high-k material. The results may, in both cases, be the decrease of the barrier for electron injection from the gate and, as a consequence, the deterioration of the erase performance. In this paper we study the effect of gate material and of the morphological transformation associated with the high-k post deposition anneal on the erase and the retention behaviour of nitride-based cells. Two different high-k dielectrics are investigated: Al_2O_3 (which has already been found to be able to significantly improve the erase operation [4, 5], guaranteeing at the same time excellent endurance and sufficient bake retention [6]) and HfAlO. We show that both for Al_2O_3 and HfAlO a trade-off exists between erase and retention, higher PDA temperatures being beneficial for erase but detrimental for retention. We also discuss the effect of Fermi level pinning and poly-Si depletion on the erase behaviour and compare the erase performances of several PVD- and AVD-deposited metal gates.

EXPERIMENTAL

Poly-Si/Al$_2$O$_3$/Si$_3$N$_4$/SiO$_2$/c-Si (SANOS) and Poly-Si/SiO$_2$/Si$_3$N$_4$/SiO$_2$/c-Si (SONOS) stacks have been fabricated using ISSG oxidation to grow the bottom SiO$_2$ layer and low-pressure CVD (LPCVD) or plasma-enhanced CVD (PECVD) processes to deposit the nitride charge trapping layer. HTO was used to deposit the top oxide in the case of the SONOS stack.

Data reported in the following, except when otherwise explicitly stated, refers to stacks with LPCVD nitride. For the SANOS stack, atomic layer deposition carried out at 350 °C (with H$_2$O as precursor) was used to deposit the Al$_2$O$_3$ films. A post deposition anneal (PDA) was carried out immediately after Al$_2$O$_3$ deposition. The Al$_2$O$_3$ thickness was varied to take into account the densification of the layer upon annealing [8] so as to achieve the target thickness of 10 nm, independently from the PDA temperature. In the following, except when explicitly reported, all the SANOS data refers to devices that received a PDA treatment of 1000 °C for 60 s in N$_2$ ambient. The thicknesses of the ANO and ONO stacks used for this work are 10/5/4 nm and 6/5/4 nm, respectively, ensuring an EOT of ≈ 12 nm for both stacks. As alternative to Al$_2$O$_3$, ALD-deposited HfAlO films with Hf atomic concentrations in the range 30%-70% have also been studied. After stack formation, either 100 nm in-situ doped poly-Si or 15 nm metal (capped with 100 nm poly-Si) was deposited. TiN and TaN were deposited using physical vapour deposition (PVD). TaC and TaCN metal gates were deposited using atomic vapour deposition (AVD). The electrical activation after poly-Si deposition was carried out at 800 °C for 30 min. The electrical characterization was performed on 50x50 μm^2 capacitors by measuring the shift of the flat band voltage (ΔV$_{fb}$) upon the program (P) or erase (E) operation with respect to the intrinsic value. All the data in this paper have been obtained using 16 V for the program and -18 V for the erase operation. Erase transients have been measured on stacks previously programmed at 16 V for 0.1 ms (ΔV$_{fb}$ ≈ + 4 V). More details on the capacitor structure and fabrication can be found in ref. [9].

RESULTS

High-k morphology

In Figure 1 the shift of V$_{fb}$ as a function of erase time is shown for SANOS and SONOS stacks. In the case of SONOS stack the erase operation saturates at +2 V. Therefore it is not even possible to reach the intrinsic level for this kind of stack. For the SANOS stack, on the other hand, the saturated flat band voltage is about -2 V, indicating that the presence of Al$_2$O$_3$ causes an increase of the P/E window of ≈ 4 V. However, the integration of Al$_2$O$_3$ is not straightforward. In particular the advantages of the Al$_2$O$_3$ layer critically depend by the morphology of the high-k layer. This is demonstrated in Fig. 2 where the erase transients after 60 s PDA anneals performed either at 1000 °C or at 700 °C are compared. The saturated V$_{fb}$ decreases by ≈ 4 V when the annealing temperature increases from 700 °C to 1000 °C. It is to be noted that, the erase performance of the 700 °C variant is even worse than on standard SONOS capacitors (Fig. 1). The advantage of Al$_2$O$_3$ over SiO$_2$ as blocking layer is therefore completely lost for low-temperature PDA treatments.

The better erase performance after the 1000 °C PDA correlates with the amorphous/crystalline phase change of the Al$_2$O$_3$ layer upon 1000 °C PDA.

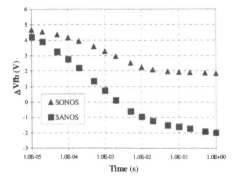

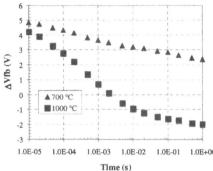

Figure 1. Shift of the flat-band voltage (ΔV_{fb}) during erase for ONO and ANO stacks (V_{gate} = -18V, p-type poly-Si gate).

Figure 2: Effect of the temperature of the Al_2O_3 PDA treatment on the erase transient of SANOS capacitors with p-type gate (V_{gate} = -18V).

In fact, in agreement with literature data [10], the film is still amorphous after the 700 °C PDA whereas, as indicated by the the plan-view TEM picture in Fig. 3a, it is fully crystallized upon the PDA treatment at 1000 °C. Selected area electron diffraction (SAED) technique shows that the crystalline phase of the film is the γ – Al_2O_3 phase (Fig. 3b).

(a)

	measured	γ - Al_2O_3	
ring	d_{hkl} (nm)	d_{hkl} (nm)	hkl
a	0.274	0.279	220
b	0.240	0.238	311
c	-	0.228	222
d	0.199	0.197	400
e	-	0.152	333,511
f	0.140	0.140	440
g	0.121	0.120	533
	0.119	0.119	622

(b)

Figure 3. (a) Plan-view TEM picture of 10 nm Al_2O_3 film after PDA annealing in N_2 ambient at 1000 °C for 60 s. The film is fully polycrystalline with crystal sizes of ~2-25nm; (b) inter-planar distances obtained from SAED pattern are in very good agreement with those characteristic of the γ-Al_2O_3 phase. The c and e rings are not observed due to texture.

Since the change in the dielectric constant of the Al_2O_3 layer upon crystallization is reported to be negligible [8], the huge impact of the PDA temperature on the erase saturation performance of the SANOS capacitors is most probably explained by the shift of the Al_2O_3 conduction band upwards by ≈ 0.5 eV upon crystallization [10]. Such a shift increases the potential barrier for FN electron tunnelling from the gate into the Al_2O_3 layer, thus hampering the root cause of the erase saturation effect. The higher temperatures associated with the crystallization of the Al_2O_3 film can however, be detrimental for data retention. This is shown in Fig. 4a, where the V_{fb} shifts after 2×10^5 s bake at 150°C measured on capacitors programmed to ΔV_{fb} of 5 V are reported. Clearly, depending on the type of nitride, a higher PDA temperature can result in deteriorated data retention. The trade off between erase and retention modulated by the PDA anneal is not only found for Al_2O_3 but also for other high-k materials like HfAlO (Fig. 4b).

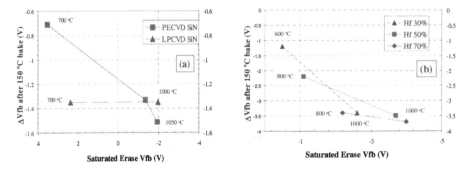

Figure 4. ΔV_{fb} loss in programmed capacitors upon bake at 150 °C versus the saturated erase V_{fb} for (a) ANO stacks with LPCVD or PECVD nitride; (b) stacks with HfAlO films deposited with different Hf atomic concentration. In both cases retention performance degrades at higher PDA temperatures. The duration of the bake was 2×10^5 s for (a) and 1×10^4 s for (b)

Poly-Si versus metal gate

Figure 5 shows that changing the poly doping from n- to p-type improves the erase saturation by only ≈ 350 mV. Spreading resistance measurements (not shown) indicate that the carrier concentration in the poly-Si layer is in the range $(1 \div 5) \times 10^{20}$ cm^{-3}. At this doping concentration the increase of the barrier for parasitic electron tunneling from the poly-Si into the nitride during the erase operation should be comparable to the Si band-gap (1.1. eV) [11]. The effect of the change of the doping polarity on the erase performance should therefore be even bigger than that observed for the Al_2O_3 amorphous/crystalline transition (Fig. 2), which is clearly not the case. CV curves (Fig.6) indicate that changing the poly doping from n- to p-type induces a shift of the work function (WF) between n- and p-type poly-Si of only 400 mV instead of the expected 1.1 V. This suggests that Fermi-level pinning [3] at the poly-Si/high-k interface is (at least) partially responsible for the observed behavior. However, even with pinning, the barrier for the parasitic tunneling from the gate during erase still is ≈ 0.4 eV higher in p-type than in n-type poly, a value comparable with that caused by the Al_2O_3 amorphous/crystalline phase change. Pinning alone is therefore not enough to explain the only marginal improvement in Fig. 5. Poly-depletion should also be taken into account. Indeed, in presence of poly-Si depletion the band bending reduces the effective barrier for parasitic tunneling from the gate (Fig. 7). For example, for carrier concentrations of ≈ 3×10^{20} cm^{-3} it can be calculated that the reduction of the barrier for electron injection during an erase operation carried out at a $V_G = -18V$ is about 0.3 eV [12]. The effect of replacing poly-Si gate with metal gate is studied in Fig. 8, which shows the ΔV_{fb} shift during erase for different metal gates. Results indicate that the saturated erase ΔV_{fb} is more than 1 V bigger for metals than for poly-Si gates, reaching for all metals values below the -3 V level required for Gigabit NAND memory applications. The erase transients look rather similar

32

for all the metals under investigation for erase times up to 10 ms, with TaC outperforming TaN, TaCN and TiN by ≈ 0.7 V for longer erase times.

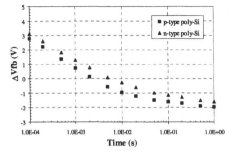 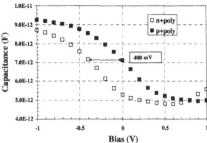

Figure 5. Shift of the flat-band voltage (ΔV_{fb}) during erase ($V_{gate} = -18V$) for ANO stacks with p- and n-type poly-Si gate gates.

Figure 6. Comparison of the CV curves for n-type and p-type poly-Si gates. The shift of the work function (WF) between n- and p-type poly-Si of only 400 mV.

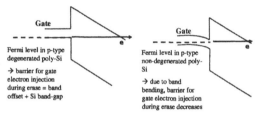

Figure 7. Schematics explaining the effect of poly-depletion on the parasitic gate electron injection during erase.

In Fig. 9 the CV curves for n- and p-type poly-Si and different metal gates are compared. Flat band voltages are similar for all metal gates. In literature TaC and TaCN deposited on HfO_2 are reported to behave as n-type and p-type materials respectively [13, 14]. The CV characteristics in Fig. 9 indicate that this is clearly not the case when these materials are deposited on Al_2O_3. Moreover, since the flat band voltage shifts of metal and p-type poly-Si gates with respect to n-type poly-Si are similar, the better erase performance for metal gates with respect to p-type poly-Si (Fig. 8) is not due to a lower Fermi level but to the absence of depletion effects in the gate.

CONCLUSIONS

In this paper we have studied the effects of gate material and high-k morphology on the erase performance of ANO stacks. We have found that: *a)* changes in the gate/high-k band offset induced by morphology changes of the Al_2O_3 layer have a strong impact on erase. In particular the shift of ≈ 0.5 eV associated with the amorphous/crystalline phase change improves the saturated erase voltage by 4 V. The high thermal budget required to trigger this transformation has however a detrimental effect on the retention, depending on the quality of the nitride layer. The trade off between erase and retention modulated by the PDA temperature has also been observed for stacks with HfAlO as top dielectric; *b)* the impact of poly-Si doping polarity on the erase performance is negligible. The cause is twofold: Fermi level pinning at the poly-Si/Al$_2$O$_3$ and poly-Si depletion; *c)* the erase performance improves if metal replaces poly-Si as gate material. The improvement is found similar for all the investigated metals (TiN, TaN, TaC and TaCN). Since both p-type poly-Si and metal gate work functions are found to be similar, the improvement is mainly caused by the absence of depletion effects in the latter case.

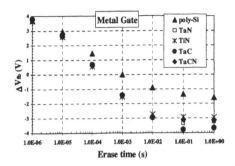 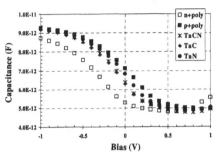

Figure 8. ΔV_{fb} shift during erase ($V_{gate} = -18V$) for PVD (TaN and TiN) and AVD (TaC and TaCN) metal gates. Erase transient for n-type poly-Si is plotted as comparison.

Figure 9. CV curve for different poly-Si and MG materials. With respect to n-type poly-Si, flat band voltage shift by the same amount ($\approx 0.4V$) both for p-type poly-Si and TaC, TaN and TaCN metal gates.

ACKNOWLEDGMENTS

This work has been done within the IMEC Industrial Affiliation Program on Advanced Flash memory together with Hynix, Infineon Technologies, Intel Corp., Micron, NXP, Samsung Electronics.

REFERENCES

1. M. White, IEEE Circuits and Devices, **16**, 22 (2000).
2. J. Van Houdt *Proceedings of the ICICDT*, p. 43 (2006).
3. C.C. Hobbs, *et al.* IEEE Trans. Electron Devices, **51**, 971 (2004).
4. C.H. Lee, *et al.*, *Proceedings of the SSDM*, p. 162 (2002).
5. C. H. Lee *et al.* – *IEDM Tech. Dig.*, p. 26.5.1 (2003).
6. Y. Shin, *et. al.*, *Proceedings of the IEEE Non-Volatile Memory workshop*, p. 58 (2003).
7. K.Park et al., *Proceedings of the IEEE Non-Volatile Memory workshop*, p. 54 (2006).
8. B. Govoreanu, *et al.* presentation AM G1.4 to the *MRS Spring Meeting* (2006).
9. A. Cacciato, *et al.*, *Proceedings of the ICMTD proceedings*, p. 217 (2007).
10. V.V. Afanas'ev, A. Stesmans, B.J. Mrstik and C. Zhao, *Appl. Phys.Lett.* **81**, 1678 (2002).
11. R. F. Pierret, in *Semiconductors fundamental- Modular series on solid state devices* (Addison-Wesley Publishing Company, 1989) p. 54.
12. A. Furnémont, PhD Thesis, University of Leuven, 2007, p. 166.
13. Y.T Hou et al. *IEDM Tech. Dig.*, (2005),
14. J. Pan et al. *IEEE Trans. Electron Devices*, **51**, 581 (2004).

Mater. Res. Soc. Symp. Proc. Vol. 1071 © 2008 Materials Research Society 1071-F02-09

Relaxation Behavior and Breakdown Mechanisms of Nanocrystals Embedded Zr-doped HfO2 High-k Thin Films for Nonvolatile Memories

Chia-Han Yang[1,2], Yue Kuo[1], Chen-Han Lin[1], Rui Wan[2], and Way Kuo[2]
[1]Texas A&M University, College Station, TX, 77843-3122
[2]University of Tennessee, Knoxville, TN, 37996

ABSTRACT

Semiconducting or metallic nanocrystals embedded high-k films have been investigated. They showed promising nonvolatile memory characteristics, such as low leakage currents, large charge storage capacities, and long retention times. Reliability of four different kinds of nanocrystals, i.e., nc- Ru, -ITO, -Si and -ZnO, embedded Zr-doped HfO_2 high-k dielectrics have been studied. All of them have higher relaxation currents than the non-embedded high-k film has. The decay rate of the relaxation current is in the order of nc-ZnO > nc-ITO > nc-Si > nc-Ru. When the relaxation currents of the nanocrystals embedded samples were fitted to the Curie-von Schweidler law, the n values were between 0.54 and 0.77, which are much lower than that of the non embedded high-k sample. The nanocrystals retain charges in two different states, i.e., deeply and loosely trapped. The ratio of these two types of charges was estimated. The charge storage capacity and holding strength are strongly influenced by the type of material of the embedded nanocrystals. The nc-ZnO embedded film holds trapped charges longer than other embedded films do. The ramp-relax result indicates that the breakdown of the embedded film came from the breakdown of the bulk high-k film. The type of nanocrystal material influences the breakdown strength.

INTRODUCTION

SiO_2 has been used as a gate oxide material in ULSIC for decades. As the minimum device dimension is scaled down from 3.5 nm to 1.5 nm, the leakage current of SiO_2 (at a gate bias of 1V) increases drastically from 10^{-12} A/cm^2 to 10 A/cm^2 due to quantum-mechanical tunneling [1]. There are many studies on replacing SiO_2 with high-permittivity (high-k) materials (e.g., Si_3N_4, $HfSi_xO_y$, HfO_2, ZrO_2) for better device performance and reliability [2]. One of the most promising high-k material applications is the high-density nonvolatile memory [3]. For the conventional poly-Si floating gate memory, any point defect in the tunnel dielectric layer can create a leakage channel to drain the stored charges to the substrate. This kind of defect path can be verified using the stress-induced leakage current experiment [5]. When the continuous poly-Si layer is replaced with nanocrystals, the above problem can be avoided [6]-[8]. In this kind of structure, discrete nanocrystalline nodes, isolated from each other by the surrounding dielectric material, can enhance the electrons or/and holes trapping capacity [3] and [4]. In early studies, nc-Si was used as the charge-storage media. Recently, various nc-metals, -metal oxides, and - semiconductors have been popular for many advantages, such as the lower power consumption, better scalability and many choices of work functions [9] and [10]. There are reports of embedding nc-Ru, -ITO, -Si, -ZnO, and -SiGe into the high-k dielectric matrix for nonvolatile memory applications [3], [11]-[13]. However, conventional high-k materials like ZrO_2 and HfO_2 crystallize at a low temperature, e.g., <600°C [3]. The Zr- or Si-doped high-k film has better bulk and interface layer properties, such as a higher crystallization temperature, a larger effective k value, and a lower interface state density, than the undoped high-k film [14]-[17]. Nanocrystals have been embedded into the Zr-doped HfO_2 (ZrHfO) dielectric in nonvolatile memories. However, there are few studies on the reliability issues of these embedded high-k films. In this paper, authors studied the relaxation behavior and the charge states of the nc-Ru, -ITO, -Si and -ZnO embedded ZrHfO dielectrics.

EXPERIMENTAL

Figure 1 shows the structure of the nanocrystals embedded high-k capacitor. All samples were deposited on the HF cleaned p-type Si (100) wafer (doping concentration at 10^{15} cm^{-3}). The ZrHfO film was deposited by reactive sputtering using a Hf/Zr (88:12 wt%) composite target in an Ar/O_2 (1:1) mixture at 5 mTorr and room temperature. Sputtering powers of the ZrHfO film and the embedded Ru, ITO, Si and ZnO layers were (100 W, 80 W), (100 W, 80 W), (100 W, 100 W), and (60 W, 60 W), respectively. For each nanocrystals embedded sample, a corresponding control sample, i.e., only the ZrHfO film without the nanocrystals embedded layer, was prepared under the same sputtering and annealing conditions. Since the nc-Ru and nc-ITO embedded films were fabricated under the same condition, only one control sample was prepared for them. The as-deposited embedded layer was amorphous, but it transformed into nanocrystals after the post-deposition annealing (PDA) step [18]. Other sample preparation conditions are shown in Table 1.

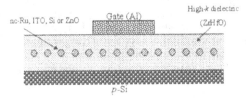

Figure 1. Cross-sectional view of a nanocrystals embedded MOS capacitor

Table 1. Embedded and non-embedded gate dielectric fabrication conditions

Sample	Post-deposition annealing temperature (°C)/gas	Post-metal annealing temperature (°C)/gas	Equivalent oxide thickness (nm)
Al/ZrHfO/nc-Ru/ZrHfO/p-Si	950/(N_2/O_2 1:1)	250/(N_2/H_2)	9
Al/ZrHfO/nc-ITO/ZrHfO/p-Si	950/(N_2/O_2 1:1)	250/(N_2/H_2)	8.6
Control Sample for nc-Ru and nc-ITO	950/(N_2/O_2 1:1)	250/(N_2/H_2)	10
Al/ZrHfO/nc-Si/ZrHfO/p-Si	950/(N_2)	300/(N_2/H_2)	10
Control Sample for nc-Si	950/(N_2)	300/(N_2/H_2)	10
Al/ZrHfO/nc-ZnO/ZrHfO/p-Si	800/(N_2)	200/(N_2/H_2)	7.8
Control Sample for nc-ZnO	800/(N_2)	200/(N_2/H_2)	6

RESULTS and DISCUSSION

The capacitance-voltage (*C-V*) characteristics of the nanocrystals embedded samples are shown in Figure 2. The *C-V* curves were measured from the accumulation region to the inversion region and back to the accumulation region in the range of (-6V, 6V, -6V). They all show counterclockwise hysteresis behavior, which means net charges trapping in the gate dielectric structure. It is obvious that the charge storage capacity is influenced by the type of the embedded nanocrystalline material. Charges injection and trapping efficiencies are related to the gate stack's physical thickness and structure. For the embedded sample, the EOT is contributed by properties of both the nanocrystal layer and the ZrHfO film. Before these two factors can be distinguished, EOT can be taken as a reference of the gate stack. Since EOTs of all samples in this study are in a close range, i.e., 8.6 nm to 10 nm except the nc-ZnO embedded (7.6 nm) and non embedded (6 nm) samples, it was assumed that the thickness effect is negligible. The charge trapping capabilities of nanocrystal embedded films are in the order of nc-ZnO>nc-Ru> nc-Si~nc-ITO. However, the charge trapping capabilities of different nanocrystal layers cannot be compared from these *C-V* hysteresis curves unless their layer thicknesses are the same. Currently, authors are making a detailed study on the charge trapping capacity of each type of nanocrystal material by taking into consideration the thickness factor. The result will be published later.

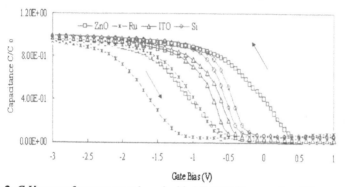

Figure 2. *C-V* curves for nanocrystals embedded ZrHfO films at -6 to 6V sweep range

Figure 3 shows the decay of the relaxation current with time (*t*) of various samples. The relaxation current (I_{relax}) of the dielectric film is the leakage current measured immediately after the sudden removal of an applied gate voltage (V_g). The I_{relax} is contributed by two co-existing mechanisms, i.e., charge trapping/detrapping and dielectric polarization/relaxation [24]. Currently, it's difficult to distinguish one from the other [19]. Each sample was stressed at a V_g of -6V for 120 seconds and the I_{relax} was measured immediately after the removal of V_g. In order to understand the influence of the stress voltage on the relaxation current, relaxation current curves from other stress conditions, i.e., $V_g = -5V$ for 120 seconds [26], are also included in these figures. Since these embedded samples were prepared from different conditions, the I_{relax}-*t* curve of each corresponding control sample is also included. The following conclusions can be summarized from Figure 3:

- a larger gate stress voltage gives a higher initial relaxation current, i.e., the initial relaxation current of nc-Ru, -ITO, -Si and -ZnO embedded film stressed at V_g = -6V is 1.18, 1.03, 1.26 and 1.31 that of the corresponding embedded film stressed at V_g = -5V.
- the initial relaxation current of a nanocrystals embedded film is larger than that of its corresponding control film.
- the decay rate of the relaxation current is dependent on the embedded nanocrystalline material.

The high I_{relax} of the embedded film is not directly related to the amount of charges stored in the film and the I_{relax} decay rate is not related to its initial I_{relax}. Charges stored in the embedded dielectric film are in deeply or loosely trapped state depending on the material properties. The I_{relax}-t curve only represents the behavior of the loosely trapped charges.

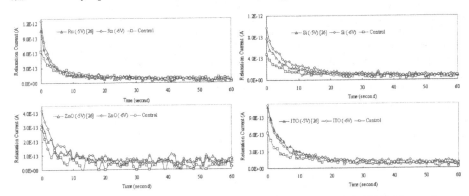

Figure 3. Relaxation current decay with time of nanocrystals embedded ZrHfO films and corresponding control samples

Figure 4 shows the log (J/P) vs. log t curves of various nanocrystals embedded and non-embedded ZrHfO (only the control film of nc-ZnO embedded sample is shown for clarity), SiO_2, HfO_x and TaO_x films [21].

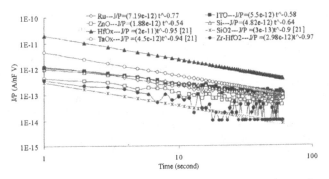

Figure 4. Relaxation current normalized to polarization vs. gate voltage release time of various embedded and non-embedded dielectrics (log-log scale)

38

The relaxation current decay rate of a dielectric layer can be expressed by the following Curie-von Schweidler equation [20]:

$$J / P = at^{-n}$$

where J is the relaxation current density (A/cm^2), P is the total polarization or surface charge density (V · nF/ cm^2), t is time in second, a is a constant, and n is a real number between 0 and 1. In Figure 4, the n values of the nanocystals embedded samples are 0.77, 0.58, 0.54 and 0.64 for nc-Ru, -ITO, -ZnO and -Si embedded samples, respectively. They are much smaller than 1 predicted in refs. [22] and [23]. The small n values indicate that the I_{relax} decays slowly. Charges trapped to the embedded nanocrystals are probably held stronger than those trapped in the bulk high-k film. Although the non-embedded ZrHfO films (control samples) were fabricated under different conditions, their charge decay rates vary little. The smaller n value of the nanocrystals embedded sample compared with that of the corresponding control sample means that for the same release time, the decrease of J/P of the former is lower than that of the latter. Based on the same P, i.e., surface charge density, the nanocrystals embedded sample has a lower relaxation current decay rate than that of the corresponding sample. Furthermore, when the small variation of the decay rates of the control samples is neglected, the charge holding strength of the nanocrystals embedded samples decreases in the order of nc-ZnO > nc-ITO > nc-Si > nc-Ru. However, strictly speaking, the relaxation behaviors of different nanocrystals embedded samples should be compared under the condition of the same nanocrystal layer thickness. More investigations are required to quantify the influence of the relaxation behavior of each type of nanocrystal material and the ZrHfO film.

By using the approximation of a two-dimensional oxide sheet charge located away from the interface, the number of initially trapped charges can be estimated by [25]

$$N = \gamma \times (C_{ox} / q) \times (- \Delta V_{FB})$$

where C_{ox} is the capacitance at accumulation region, q is the electron charge, ΔV_{FB} is the flatband voltage shift, and γ is a correction factor that can be estimated by the ratio of the sum of control and tunnel oxide EOT to the control oxide EOT [6]. Equations shown in Figure 4 can be used to calculate the number of loosely trapped charges released in a period of time. The difference between the number of the initially trapped charges and that released from the relaxation current could be considered as the number of deeply trapped charges in the dielectric structure. Table 2 shows the percentages of deeply and loosely trapped charges of 4 kinds of embedded samples within 20 seconds after releasing of the applied gate voltage. Less than half of the initially trapped charges were released from the relaxation process. Therefore, the charge holding capability of the nanocrystal is in the order of nc-ZnO > nc-Si > nc-Ru > nc-ITO. Moreover, this result shows that some of the difference in hysteresis of each nc-embedded capacitor in Figure 2 may be attributed to the difference in their EOTs. For example, from C-V curves of nc-Si and nc-Ru embedded samples, nc-Ru embedded film can trap much more charges, but nc-Si has much stronger charge holding capability than nc-Ru. That is, the large hysteresis in nc-Ru embedded films cannot guarantee a strong charge holding strength.

Ramp-relax method, proposed in ref. [23], has been demonstrated can effectively monitor the breakdown mechanism of high-k thin films. In this experiment, a negative gate voltage (-V$_g$)

was applied to the nanocrystals embedded capacitor. Then, the leakage current density (J_{ramp}) was measured. Subsequently, the applied voltage was released for a short time followed by immediate measurement of J_{relax}. The above procedure was repeated until the dielectric layer broke at a large -V_g, i.e., abrupt increase of J_{ramp}. Before the high-k film was broken, the polarity of the J_{relax} should be opposite to that of the J_{ramp} because of the polar structure of the high-k film; after the high-k stack was totally broken, the J_{relax} would show the same polarity as the J_{ramp} [27].

Table 2. Percentages of deeply and loosely trapped charges for nanocrystal embedded samples after the first 20 seconds.

nanocrystal	deeply trapped %	loosely trapped %
Ru	78.5%	21.5%
ITO	57.5%	42.5%
Si	85.8%	14.2%
ZnO	93.6%	6.4%

Figure 5 shows the J_{ramp}-Vg and J_{relax}-Vg curves of these capacitors. Each J_{ramp}-Vg curve contains three sections:

- Pre-breakdown: J_{ramp} increases with -V_g very slightly and smoothly because charges begin to stack in the bulk high-k to form a spot-connected breakdown path [28].
- Quasi-breakdown: J_{ramp} increases faster due to the formation of a small number of connected paths [28].
- Complete breakdown: J_{ramp} jumps abruptly to a very large number because the film becomes conductive.

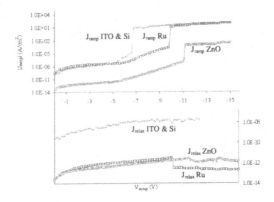

Figure 5. Ramp-relax test results on nanocrystals embedded high-k films

However, the J_{relax} did not show the same polarity as the J_{ramp} after the sample is broken, which is different from the breakdown phenomenon of the non-embedded high-k film [27]. The embedded nanocrystals are metals or semiconductors, which are very difficult to breakdown, so the failure of the embedded film must be due to the breakdown of the bulk ZrHfO film. The existence of the relaxation current with the same polarity as before breakdown is due to the strong hold of the trapped charges on the embedded nanocrystals. Figure 5 also shows that the breakdown voltage is dependent to the embedded material, i.e., in the order of nc-ITO ~ nc-Si < nc-Ru < nc-ZnO. Since the physical thickness and the distribution of nanocrystals in the high-k film vary with the material properties of the embedded material and the process condition, it is difficult to understand the exact influence of the nanocrystal material on the breakdown strength. More studies are required.

CONCLUSIONS

Large C-V hystereses were detected in four different kinds of nanocrystals, i.e., nc-Si, -ITO, -Ru, and -ZnO, embedded ZrHfO high-k films. It is difficult to compare the charge storage capacities of these samples because they have different physical thicknesses or EOTs. Relaxation currents of the nanocrystals embedded high-k dielectric films were measured and compared. A larger gate voltage provides a higher initial relaxation current than the lower gate voltage does. The relaxation behavior does not seem to be influenced by the initial applied gate voltage. Different nanocrystals held charges at different strengths, i.e., in the order of nc-ZnO > nc-Si > nc-Ru > nc-ITO. In order to understand the charge storage process of the nanocrystals embedded film, both the C-V hysteresis and the relaxation behavior have to be investigated; the former is used to calculate the total number of initially trapped charges and the latter is for the estimation of the number of loosely trapped charges. The relaxation current was contributed by the loosely trapped charges, which is less than half of the originally trapped charges. The nanocrystal material influences not only the charge storage capacity but also the charge holding time. The breakdown of the nanocrystals embedded sample was due to the breakdown of the bulk high-k film instead of the nanocrystals, which was demonstrated from the lack of polarity change of the relaxation current.

ACKNOWLEDGMENTS

This project was partially supported by NSF CMMI-0654172. Authors acknowledge Dr. J. Lu for preparing the nc-Si and nc-ZnO embedded ZrHfO capacitors.

REFERENCES

1. The International Technology Roadmap for Semiconductors. Semiconductor Industry Association, December (2003).
2. S. Chatterjee, S. K. Samanta, H. D. Banerjee and C. K. Maiti, Semicond. Sci. Technol., 17, 993 (2003).
3. J. Lu, Y. Kuo, J. Yan and C.-H. Lin, Jpn. J. Appl. Phys., 45 (34), L901 (2006).
4. Y. Kuo, J. Yan and C.-H. Lin, in Proceedings of the 6th IEEE conference on Nanotechnology, 469 (2006).

5. N. Mielke and J. Chen, in *Oxide Reliability: A Summary of Silicon Oxide Wearout, Breakdown and Reliability*, D. J. Dumin, 103 (2002).
6. S. Tiwari, F. Rana, H. Hanafi, A. Hartstein, E. F. Crabbé, and K. Chan, Appl. Phys. Lett., **68** (10), 1377 (1996).
7. B. De Salvo, G. Ghibaudo, G. Pananakakis, P. Masson, T. Baron, N. Buffet, A. Fernandes and B. Guillaumot, IEEE Trans. on Electron Devices, **48**, 1789 (2001).
8. J. De Blauwe, IEEE Trans. Nanotechnol., **1**, 72 (2002).
9. Z. Liu, C. Lee, V. Narayanan, G. Pei and E. C. Kan, IEEE Trans. Electron Devices, **49**, 1606 (2002).
10. Z. Liu, C. Lee, V. Narayanan, G. Pei and E. C. Kan, IEEE Trans. Electron Devices, **49**, 1614 (2002).
11. J. Lu, C.-H. Lin and Y. Kuo, ECS Trans., **11** (4), 509 (2007).
12. A. Birge and Y. Kuo, Journal of the Electrochemical Society, **154** (10), H887 (2007).
13. D. B. Farmer and R. G. Gordon, J. Appl. Phys., **101**, 124503 (2006).
14. Y. Kuo, J. Lu, S. Chatterjee, J. Yan, T. Yuan, H.-C. Kim, W. Luo, J. Peterson and M. Gardner, in ECS. Trans., **1**, 447 (2006).
15. D. Triyoso, ECS Trans., **3**, 463 (2006).
16. Y. Kuo, ECS. Trans., **3**, 253 (2006).
17. Y. Kuo, ECS. Trans., **2**, 13 (2006).
18. J. Yan, Y. Kuo and J. Lu, Electrochem. Solid-State Lett., **10** (7), H199 (2007).
19. J. R. Jameson, W. Harrison, P. B. Griffin and J. D. Plummer, Appl. Phys. Lett., **84**, 3489 (2004).
20. A. K. Jonscher, *Dielectric Relaxation in Solids*. New York: Chelsea (1983).
21. H. Reisinger et al., in *IEDM Tech. Dig.*, 267 (2001).
22. Z. Xu, L. Pantisano, A. Kerber, R. Degraeve, E. Cartier, S. De Gendt, M. Heyns and G. Groeseneken, IEEE Trans. on Electron Devices, **51** (3), 402 (2004).
23. W. Luo, Y. Kuo and W. Kuo, IEEE Trans. on Device and Materials Reliability, **4** (3), 488 (2004).
24. W. Luo, T. Yuan, Y. Kuo, J. Lu, J. Yan and W. Kuo, Appl. Phys. Lett., **89**, 072901 (2006).
25. E. H. Nicollian and J. R. Brews, *Metal Oxide Semiconductor Physics and Technology*, 478, Wiley, Hoboken, NJ (2003).
26. C.-H. Yang, Y. Kuo, R. Wan, C.-H. Lin and W. Kuo, "Failure Analysis of Nanocrystals Embedded High-k Dielectrics for Nonvolatile Memories," submitted to IEEE International Reliability Physics Symposium (2008).
27. W. Luo, T. Yuan, Y. Kuo, J. Lu, J. Yan and W. Kuo, Appl. Phys. Lett., **88**, 202904 (2006).
28. H. Satake and A. Toriumi, IEEE Trans. on Electron Devices, **47** (4), 741 (2000).

Mater. Res. Soc. Symp. Proc. Vol. 1071 © 2008 Materials Research Society 1071-F02-11

Simulation of Capacitance-Voltage Characteristics of Ultra-Thin Metal-Oxide-Semiconductor Structures With Embedded Nanocrystals

Mosur Rahman[1], Bo Lojek[2], and Thottam Kalkur[3]
[1]University of Colorado at Colorado Springs, Colorado Springs, CO, 80933-7150
[2]Atmel Corporation, Colorado Springs, CO, 80907
[3]Electrical and Computer Engineering Dept., University of Colorado at Colorado Springs, Colorado Springs, CO, 80933-7150

Abstract

This paper presents an approach to model and simulate quantum mechanical (QM) effects in solid-state devices such as Metal Oxide Semiconductor (MOS) capacitor with and without nanocrystal in the oxide. This QM model is developed to understand finite inversion layer width and threshold voltage shift. It allows a consistent determination of the physical oxide thickness based on an agreement between the measured and modeled CV curves. However, as for thinner oxides finite inversion layer width effects become more severe, QM model predicts higher threshold voltage than the classical model. The inversion-layer charge density and MOS capacitance in multidimensional MOS structures are simulated with various substrate doping profiles and gate bias voltages. The effectiveness of the QM correction method for modeling quantum effects in ultrathin oxide MOS structures is also investigated. The CV characteristic is used as a tool to compare results of the Schrödinger–Poisson (SP) solution i.e. the QM model with that of the QM correction, the Classical solution and measured data. The change in C-V characteristics indicative of threshold voltage shift for Si nanocrystal embedded in oxide is also investigated.

I. INTRODUCTION

Quantum mechanics have played a significant role primarily in compound semiconductor devices. However, due to the shrinking feature size of CMOS devices toward tens of nanometers in gate length, the QM effects manifest even in the conventional silicon devices [1-4]. The electrical properties of nanoscale semiconductor devices and structures are typically determined by quantum confinement that strongly affects the density of states of electrons and holes. When energy bands are bent strongly near a semiconductor-insulator interface a potential well formed by the interface barrier and the electrostatic potential in the semiconductor can be narrow enough that quantum-mechanical effects become important. So the shrinking dimensions of the devices require suitable device models in view of physics and mathematics for accurate simulation. Since the operation of the nanoscale MOS is primarily based on controlling the electron density by varying the confining potential, the modeling of the potential distribution and the electronic states is important. Using an initial value calculated from charge neutrality for potential, potential and charge density profiles in equilibrium are computed by solving the set of nonlinear partial differential equations described by the Poisson's equation. The electronic states in the quantum dot are subsequently determined from solutions of the Schrödinger equation using the previously calculated confining potential.

A polycrystalline-oxide-silicon MOS structure with a p-type silicon substrate as shown in Fig.1 [2] is simulated. An ideal oxide with a dielectric constant of 3.9 and a Si substrate with a dielectric constant of 11.7 are assumed. The oxide–silicon interface is chosen to be at x=0. The polysilicon (or metal) work function, oxide thickness, substrate doping concentration and temperature can be varied. The silicon layer lies in the region and the doping concentration of silicon is assumed to be 1e17 per cm^3.

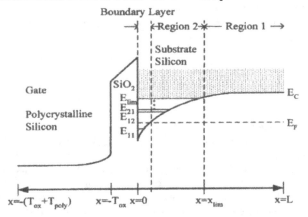

Fig. 1. A schematic band profile for the MOS structure.

QM effects in the surface due to potential well have a profound impact on both the amount and profile of charge. The critical dimension in traverse direction is the gate oxide thickness, which now has fallen below 2 nm for the most advanced MOS devices. To extract information about the charge distribution alone requires the solution of Schrödinger and Poisson equations, assuming QM effects are to be fully accounted for.

II. SIMULATION PROCEDURES

As mentioned earlier when electrons are confined, the equilibrium electrical characteristics of semiconductor devices are modeled by solving the coupled Schrödinger and Poisson equations self-consistently. We have to deal with an elliptic system of partial differential equations that reads in [5]:

Poisson equation:
$$\nabla^2 \phi = -\frac{q}{\varepsilon}\left(p - n + N_D^+ - N_A^- + \rho_F\right)$$

and Schrödinger equation:
$$-\frac{\hbar^2}{2m}\nabla^2 \psi + U\psi = E\psi$$

where $U = -q\phi + \Delta E_c$ and

$$n = \sum_i^{All\ Bound\ States} \frac{m^* k_B T}{\pi \hbar^2} \ln\left[1 + \exp\frac{(E_f - E_i)}{k_B T}\right]|\psi_i|^2 \qquad \text{for 1D}$$

$$n = \sum_i^{All\ Bound\ States} \frac{\sqrt{2m^* k_B T}}{\pi \hbar} F_{-1/2}\left[\frac{(E_f - E_i)}{k_B T}\right]|\psi_i|^2 \qquad \text{for 2D}$$

44

$$n = 12\sum_{i}^{l} |\psi_i|^2 + m|\psi_{l+1}|^2$$ with total number of electron, $N = 12 \times l + m$ $(m < 12)$ for 3D

Where ψ = wave function of the carrier, $\hbar$ = reduced Planck constant = $\dfrac{h}{2\pi}$, m = electronic mass, E_c = conduction band in nanocrystal, k_B = Boltzmann constant, E_f = Fermi energy, E_i = i-th energy level, T = temperature

The primary function of our code is to solve these partial differential equations self-consistently for the electronic potential, ϕ in volts, and the electron and hole concentration, n and p in cm^{-3} respectively.

The Poisson equation is discretized with finite difference method and solved by Incomplete Complex Conjugate Gradient (ICCG) method. This iterative method is based on the conjugate gradient algorithm applied to the implicitly formed normal equations with a pre-conditioner to accelerate convergence. The pre-conditioner is based on an incomplete LU factorization. The Schrödinger equation is discretized using finite difference method and solved by using ARnoldi PACKage (ARPACK). In the finite difference method the differential equations at the inner points are replaced by difference equations where only the nearest neighboring points for each of the inner points are invoked. The program ends when the difference between the two potentials at the end of two successive cycles is lower then a fixed tolerance.

The differential channel capacitance for a change in the gate potential of a MOS structure is approximated from the Q-V data using finite difference method where Qs(Vg) is the total charge in the channel region at a given gate bias, Vg.

$$Cs = \frac{dQs}{dVg} = \frac{Qs(Vg_1) - Qs(Vg_2)}{Vg_1 - Vg_2}$$

III. QM MODELING OF SILICON MOS DEVICES:

The channel charge is altered from its classical distribution by QM effects through two mechanisms [3]:

1) The quantization of energy bands in the potential well effectively raises the ground energy level available for carrier to occupy in the surface region;

2) The carrier density distribution in the transverse direction is determined by the superimposition of wavefunctions at discrete energy levels, both of which can be obtained from solving the Schrödinger equation in the surface potential well and applying the Fermi–Dirac statistics.

A zero wave function boundary condition is forced at quantum system boundaries, i.e., at the bottom of the substrate and at the oxide interface. Due to the repulsive boundary condition at the Si/SiO$_2$ interface to the wavefunctions, the resulting carrier profile peaks at a certain distance away from the interface in the surface quantum.

There have been two approaches to the modeling of these quantum effects: 1) employing full quantum mechanical transport model such as nonequilibrium Green's function [5] and 2) adding quantum corrections to the classical drift-diffusion (D-D) or hydrodynamic equation, such as the density gradient method [6]. The QM correctional analytical model proposed by Hansch [7], considers the repulsive boundary condition for channel carriers at the Si/SiO$_2$ interface. To satisfy this boundary condition, the Hansch model introduces a shape function, which is to be imposed upon carrier concentration in the transverse direction. Effectively, the 3-D density of states becomes a function of depth with near zero value at the surface. Device simulation incorporating the Hansch model gives carrier profile as shown in Fig. 2.

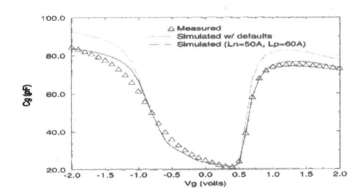

Fig. 2. Simulated C-V characterictics using Hansch's model compared to measured data for a MOS capacitor with gate oxide thickness of 31 Å.

The shortcoming is that the Hansch model neglects the fact that the ground energy level is raised to above the band edge due to the energy quantization, which has a direct impact on the threshold voltage shift. Fig.2 compares the simulation results with the measured data for an nMOS capacitor (p-substrate) with the gate oxide thickness of 31 Å using classical and Hansch models. The Hansch model fails to explain the larger capacitances in different gate biases mode. To achieve better simulation accuracy, the values for key parameters have to be adjusted from their physically meaningful default values.

IV. SIMULATION RESULTS AND DISCUSSION

We present the calculated CV using SP and classical models. From the solution, there are two prominent distinct features in comparison.
1) The channel carriers now are distributed among discrete eigen energy levels instead of in a single energy band.
2) The peak of the space carrier concentration is located some distance away from the surface in the substrate.

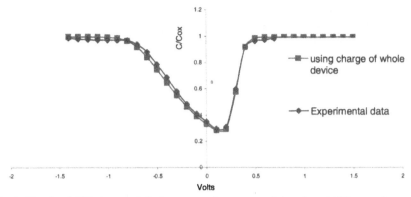

Fig. 3. Simulated CV characteristic is compared to measured data for a MOS capacitor with gate oxide thickness of 20 Å.

The ratio of total capacitance and oxide capacitance is plotted for variation in applied gate voltage as shown in Fig. 3. Here experimental data [3] is compared to the simulated result for a MOS structure with 2 nm gate oxide. To estimate the effective charge we have used different approaches in Fig. 4. Estimating charge using fixed depth of channel (1nm) gives almost identical result to depth defined by a change in potential of a predefined value, 50mV. In this case the change in potential is large enough to account for only the charge close to surface and channel. Downward spike occurred at the flat band voltage in the second case due to numerical constraint when zero charge presence was detected. When the whole substrate is considered for charge calculations, the CV curves flattened out due to the increased channel depth. When a small change in potential is used to calculate the channel depth, it essentially takes the whole device.

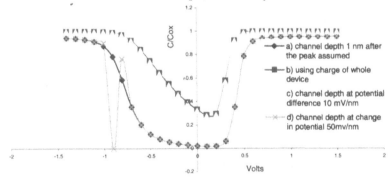

Fig. 4: Simulation of CV characteristics for a p type MOS with channel depth define by a) fixed b) whole device c) potential difference greater that 10 mV/nm d) greater than 50 mV/nm.

When a cubic nano-crystal is placed in the oxide, the threshold voltage increases as higher voltage is required to get the same inversion. Comparison of CV characteristics of MOS structure with and without nano-crystal is shown in Fig. 5. For negative gate voltage in the depletion mode the CV characteristic curve is essentially the same. In this mode charge consists of only holes [1]. From the inversion region CV characteristics of MOS with a nano-crystal differs from the characteristics without a nano-crystal, as a nano-crystal changes the potential profile in the semiconductor, which in turns alter the charge distribution profile in the channel. To get rid of the numerical uncertainty at the flat band voltage, an average value is assigned for the capacitance at that voltage.

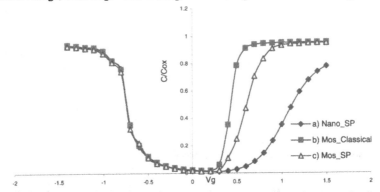

Fig. 5: Simulation of CV characteristics for a p type MOS with a) nano-crystal using SP method (Nano_SP) b) without nano-crystal using classical (Mos_Classical) and c) without nano-crystal using SP method (Mos_SP).

IV. CONCLUSION

In this paper, we have investigated quantum effects occurring at the oxide–silicon interface under the inversion condition with the SP method for the ultra thin oxide MOS structure. The range of the optimum parameter is determined for various doping profiles at different gate voltages. Classical and QM corrections introduce an error in the C–V measurement curves. C-V characteristics for MOS structure with and without embedded nanocrystals are analyzed. Presence of nanocrystals increased the threshold voltage of MOS structure as predicted. This shift depends on the size, shape and doping of the nanocrystals.

References

[1] S.-H. Lo, D. A. Buchanan and Y. Taur, "Modeling and characterization of quantization, polysilicon depletion, and direct tunneling effects in MOSFETs with ultrathin oxides", IBM J. Res. Develop. Vol. 43 No. 3, pp. 327-337 May 1999
[2] Yiming Li, Ting-wei Tang and Xinlin Wang, "Modeling of Quantum Effects for Ultrathin Oxide MOS Structures With an Effective Potential", IEEE Transaction on Nanotechnology, Vol. 1, No. 4, pp. 238-242, December 2002
[3] Zhiping Yu, Robert W. Dutton and Richard A. Kiehl, "Circuit/Device Modeling at the Quantum Level", IEEE Transaction on Electron Devices, Vol. 47, No. 10, pp. 1819-1825, October 2000
[4] V. Y. Thean, S. Nagaraja and J. P. Leburton, "Three-dimensional self-consistent simulation of interface and dopant disorders in delta-doped grid-gate quantum-dot devices", J. Appl. Phys. 82(4), pp. 1678-1686, August 1997
[5] R. Lake, G. Klimeck, R. C. Bowen and D. Jovanovic, "Single and multiband modeling of quantum electron transport through layered semiconductor devices", J. Appl. Phys. Vol. 81, No. 12, pp 7845-7869, June 1997
[6] M. G. Ancona, "Equations for state for silicon inversion layers", IEEE Transaction on Electron Devices, Vol. 47, pp 1449-1456, July 2000
[7] W. Hansch, T. Vogelsang, R. Kircher and M. Orlowski, "Carrier transport near the Si/SiO2 interface of MOSFET", Solid-State Electron., Vol. 32, pp.839, 1989

Mater. Res. Soc. Symp. Proc. Vol. 1071 © 2008 Materials Research Society   1071-F03-16

Synthesis of Nickel Disilicide Nanocrystal Monolayers for Nonvolatile Memory Applications

Jong-Hwan Yoon[1], and Robert G. Elliman[2]

[1]Department of Physics, College of Natural Sciences, Kangwon National University, Chuncheon, Gangwon-do, 200-701, Korea, Republic of

[2]Department of Electronic Materials Engineering, Research School of Physical Science and Engineering, The Australian National University, Canberra, ACT, 0200, Australia

ABSTRACT

Nickel silicide nanocrystals (NCs) were formed by thermally annealing SiO_xN_y films either implanted with Ni or coated with an evaporated Ni film. It is observed that the NCs grow into well-defined single crystalline structures embedded in a SiO_xN_y matrix, and that their size can be directly controlled by adjusting the concentrations of either silicon or nickel in the SiO_xN_y layer. The formation of well-defined NC monolayers was also demonstrated by depositing an ultra-thin Ni layer between two SiO_xN_y layers. These structures are shown to exhibit characteristic capacitance-voltage hysteresis suitable for nonvolatile memory applications.

INTRODUCTION

The demand for nonvolatile memory (NVM) devices, such as memory sticks in digital cameras, is rising rapidly and as a consequence there is considerable research effort devoted to realizing devices with smaller size, faster operating speed and larger storage capacity. One such approach is to use a floating-gate transistor where the floating-gate consists of discrete charge traps (nanocrystals) instead of a continuous conducting layer as used in many conventional devices [1].

Discrete charge storage offers many advantages over conventional floating gate structures. The charge trapped in a nanocrystal floating gate is more stable than in a conventional floating gate as the latter can lose trapped charge through a single leakage path in the gate oxide. As a consequence nanocrystal floating gate memory is expected to have a longer retention time than the conventional devices. The reduced susceptibility to gate oxide failure also means that devices can be scaled to smaller dimensions by reducing the tunnel oxide thickness.

The properties of NVM with floating gates based on semiconducting [2-4] or metallic nanocrystals [5-7] have been extensively studied. A comparison of these studies suggests that there are additional advantages in using metallic nanocrystals instead of semiconducting nanocrystals, namely, a reduction in spurious effects caused by traps at the nanocrystals/oxide interface and an enhancement in charge storage capacity and retention time. As a result, floating gates based on various metal nanocrystals, such as Au, Ag, Pt [8] and Co [9], have received particular attention. As metal silicides have been shown to have physical properties similar to those of pure metals [10] they are also clearly worthy of investigation. In this study we report the direct growth of crystalline nickel silicide nanocrystal monolayers (NCs) in silicon-rich silicon oxide (SiO_xN_y) layers and the memory properties of Ni-based nanocrystal floating gate structures.

EXPERIMENTAL DETAILS

Nickel silicide NCs were formed by thermally annealing Ni implanted/coated silicon-rich silicon oxide (SiO_xN_y) films. SiO_xN_y films were deposited on (100) Czochralski silicon wafers at a substrate temperature of 300 °C by plasma-enhanced chemical vapor deposition (PECVD) using a fixed flow rates of SiH_4 and N_2O. The Ni implanted samples were prepared by irradiating these films with 130 keV Ni^+ ions to a fluence of $6.0x10^{16}$ cm^{-2}, which nominally produces a peak Ni concentration of 10.0 at. %. The Ni coated samples were prepared by depositing an ultra-thin Ni layer between two SiO_xN_y layers. Nucleation and growth of Ni silicide NCs was achieved by annealing the Ni implanted/coated samples at 1100 °C in a quartz-tube furnace using high purity nitrogen gas (99.999 %) as an ambient. The microscopic structure of the annealed SiO_xN_y layers, and the size and crystallinity of NCs were studied by transmission electron microscopy (TEM) using a JEOL JEM 2010 instrument operating at 200 kV. The chemical composition of NCs was analyzed by energy dispersive X-ray spectroscopy (EDS) using an energy dispersive spectrometer attached to the TEM instrument. For the EDS analysis, the electron beam was focused to a spot as small as 1.5 nm in size. Capacitance-voltage measurements were also performed on selected structures formed by the Ni coated method in order to explore their potential for memory applications.

RESULTS AND DISCUSSION

Formation of $NiSi_2$ nanocrystals by nickel implantation

Figure 1 shows the simulated depth profile of Ni implanted SiO_2 and cross-sectional transmission electron micrographs (XTEM) of Ni-implanted SiO_xN_y thin films (~200 nm) with different Si concentrations after thermal annealing at 1100 °C for 4 hours.

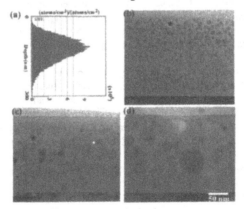

Figure 1. a) Depth profile of nickel implanted into SiO_xN_y and cross-sectional transmission electron microscopic images of b) $SiO_{1.42}N_{0.25}$, c) $SiO_{1.16}N_{0.32}$, and d) $SiO_{0.85}N_{0.35}$ samples.

The initial Ni distribution is shown to be approximately Gaussian with a mean projected range of 100 nm and a peak Ni concentration of 10.0 at. %. After annealing the layers contain a large number of precipitates with approximately spherical shape, as shown in the XTEM images of Figs 4b-d. The size of the precipitates varies for different SiO_xN_y film compositions, generally increasing with increasing Si concentration. We have previously shown that the size of the precipitates is also influenced by the supply of Ni [11]. These results demonstrate that the size of precipitates can be influenced by varying either the Si or Ni concentration in the SiO_xN_y layer.

Phase identification of the precipitates observed in the present samples was explored by electron diffraction and energy dispersive X-ray spectroscopy (EDS). Figs. 2a and 2b show a representative electron diffraction pattern and EDS spectrum taken from the sample shown in Fig. 1c. Indexing of the diffraction pattern showed that the rings were consistent with the {111}, {200}, {220} and {311} planes of $NiSi_2$. The EDS data also show that the Ni:Si ratio is about 1:2. This clearly demonstrates that the precipitates contain Si and Ni, supporting the premise that they are nickel disilicide ($NiSi_2$) nanocrystals. Details of this analysis are described in our previous work [11].

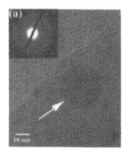 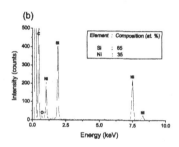

Figure 2. a) HRTEM and electron diffraction pattern of the sample in Fig. 1c, and b) EDS spectrum of an individual precipitate.

Formation of $NiSi_2$ nanocrystals by nickel coating

The formation of Ni silicide NCs was also explored by thermal annealing samples with an ultra-thin Ni layer sandwiched between two SiO_xN_y layers. Fig. 3 shows XTEM images of SiO_xN_y layers with various Si concentrations and Ni layer thicknesses after annealing at 1100 °C for 1 hour. Fig. 3a shows images from a sample consisting of three Ni layers in the form SiO_xN_y /Ni/ SiO_xN_y. In this case the composition of each SiO_xN_y layer was varied, with x=1.09, x=1.33, and x=1.77, while the thickness of the Ni layers were held constant at a thickness of 0.5 nm. In contrast, Fig. 3b shows an XTEM image of a SiO_xN_y (x=1.77) sample with three different Ni-layer thicknesses (0.5nm, 0.8nm and 0.3nm). It is clear from these data that the size of the nickel-based NCs increases with increasing Si concentration and Ni thickness, consistent with those observed in the Ni-implanted SiO_xN_y layers shown in Fig. 1. However, the Ni-coated SiO_xN_y

layers produce a much narrower spatial distribution of nanocrystals thereby demonstrating the efficacy of this technique for making nanocrystal floating gate structures for nonvolatile memory devices.

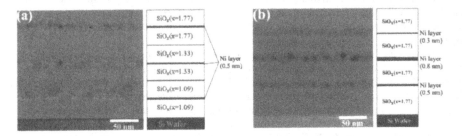

Figure 3. XTEM images of a) sample prepared by depositing a 0.5nm Ni layer between SiO_xN_y layers of composition, x=1.09, 1.33 and 1.77 and b) sample prepared by depositing different Ni thickness between SiO_xN_y layers with x=1.77.

Memory effects of NiSi$_2$ nanocrystals

The capacitance-voltage (C-V) characteristics of structures containing Ni-based NC monolayers were examined. Two MOS structures, one with and one without Ni-based nanocrystals in the floating gate were prepared for comparison.

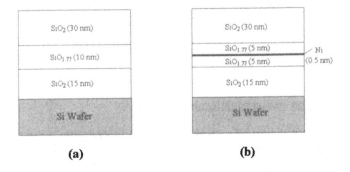

Figure 4. Schematics of MOS structures employed for capacitance-voltage measurements: a) without and b) with Ni-based nanocrystals.

Fig. 4a shows the structure of the sample without nanocrystals which was prepared by depositing a SiO_2 layer of 15 nm, a SiO_xN_y layer with x=1.77 of 10 nm, and a SiO_2 layer of 30 nm onto a p-type Si substrate. Fig. 4b shows structure of the sample with Ni-based nanocrystals which was prepared by depositing a SiO_2 layer of 15 nm, a SiO_xN_y with x=1.77 of 5 nm, a Ni layer of 0.2nm, a SiO_xN_y with x=1.77 of 5 nm and a SiO_2 layer of 30 nm onto a p-type Si substrate. The two samples were annealed at 900 °C for 1 hour. Circular Al gate electrodes of 0.9 mm^2 were subsequently formed by evaporating Al through a mask onto the sample surface to form MOS structures. The results of C-V measurements are shown in Fig. 5 which shows high-frequency (1 MHz) capacitance-voltage characteristics obtained from the MOS structures shown in Fig. 4.

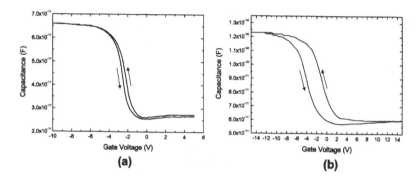

(a) (b)

Figure 5. Capacitance-voltage characteristics of MOS structures: a) without and b) with Ni-based nanocrystals.

The gate voltage was swept from negative to positive values before being swept back from positive to negative values. Both samples exhibit hysteresis loops, indicating the presence of charge traps in the floating gate. The hysteresis in Fig. 5a is likely to be associated with defects and/or Si nanocrystals, while the much greater hysteresis in Fig. 5b is believed to be associated with Ni-based NCs. There is clearly a distinct difference in the memory window for the two samples, with windows of 0.3 eV and 2.5 eV estimated for the samples in Fig. 5a and 5b, respectively. The memory properties are clearly improved by the presence of the Ni-based nanocrystals.

CONCLUSIONS

Nickel silicide nanocrystals (NCs) were formed by thermal annealing of SiO_xN_y films either implanted with Ni or coated with an evaporated Ni film. It was observed that Ni-based NCs grow into well-defined single crystalline structures embedded in the SiO_xN_y matrix, and that their size can be directly controlled by adjusting the concentrations of either silicon or nickel. The formation of a well-defined Ni-based NC monolayers was also demonstrated by depositing an ultra-thin Ni layer between two SiO_xN_y layers. These latter structures are shown to exhibit characteristic capacitance-voltage hysteresis with a memory window suitable for non-volatile memory applications.

ACKNOWLEDGMENTS

This work was supported by the Korea Research Foundation Grant funded by the Korean Government (MOEHRD, KRF-2007-313-C00269) and by the Australian Research Council through its Discovery Grant Program.

REFERENCES

1. D. Kahng and S. M. Sze, *J. Bell Syst. Tech.* **46**, 1283 (1967).
2. S. Tiwari, F. Rana, K. Chan, L. Shi, and H. Hanafi, *Appl. Phys. Lett.* **69**, 1232 (1996).
3. J. Wang, L. Wu, K. Chen, L. Yu, X. Wang, J. Song, and X. Huang, *J. Appl. Phys.* **101**, 014325 (2007).
4. C. H. Tu, T. C. Chang, and P. T. Liu, H. C. Liu, S. M. Sze, C. Y. Chang, *Appl. Phys. Lett.* **89**, 162105 (2006).
5. Z. Liu, C. Lee, V. Narayanan, C. Pei, and E. C. Kan, *IEEE Trans. Electron Dev.* **49**, 1606 (2002).
6. Z. Liu, C. Lee, V. Narayanan, C. Pei, and E. C. Kan, *IEEE Trans. Electron Dev.* **49**, 1614 (2002).
7. J. J. Lee and D. L. Kwong, *IEEE Trans. Electron Dev.* **52**, 507 (2005).
8. C. Lee, J. Meteer, V. Narayanan, and E. Kan, *J. Electron Meter.* **34**, 1 (2005).
9. D. Zhao Y. Zhu, and J. Liu, *Solid State Electronics* **50**, 268 (2006).
10. J. Yuan, G. Z. Pan, Y. L. Chao, and J. C. S. Woo in *Nickel Silicide Work Function Tuning Study in Metal-Gate CMOS Applications*, edited by G. J. Brown, R. M. Biefeld, C. Gmachl, M. O. Manasreh and K. Unterrainer, (Mater. Res. Soc. Proc. **829**, San Francisco, CA, 2005) pp. B7.11.1-B7.11.6.
11. J. H. Yoon and R. G. Elliman, *J. Appl. Phys.* **99**, 116106 (2006).

Mater. Res. Soc. Symp. Proc. Vol. 1071 © 2008 Materials Research Society   1071-F03-21

Using Thermal Oxidation and Rapid Thermal Annealing on Polycrystalline-SiGe for Ge Nanocrystals

Chyuan-Haur Kao[1], C. S. Lai[1], M. C. Tsai[1], C. H. Lee[1], C. S. Huang[1], and C. R. Chen[2]

[1]Electronics Engineering, Chang Gung University, 259 Wen-Hwa 1st Road, Kwei-Shan, Tao-Yuan, 333, Taiwan

[2]Material Science Service Corporation, Hsin Chu, 300, Taiwan

ABSTRACT

In this paper, simple techniques were proposed to fabricate germanium nanocrystal capacitors by one-step thermal oxidation and/or rapid thermal annealing on polycrystalline-SiGe (poly-SiGe) deposited with a LPCVD (low pressure chemical vapor deposition) system. This thermal oxidation method can directly result in the top-control oxide layer via the oxidation of amorphous-Si film and the formation of Ge nanocrystals from the poly-SiGe film. Otherwise, the rapid thermal annealing method can be also used to form Ge nanocrystals as comparison.

INTRODUCTION

The first NC Flash memory device was demonstrated using Si-NC embedded in SiO_2 [1]. Since then, various materials such as Ge and metals have been used to form NC FG on SiO_2 [2-3] and various storage mechanisms have been proposed. Therefore, silicon (Si) or Germanium (Ge) nanocrystals are promising candidates for flash electrically erasable and programmable read only memories (EEPROMs), and the reduction of charge leakage from weak spots in tunneling oxide.

In this work, simple techniques were proposed to fabricate germanium nanocrystal capacitors by one-step thermal oxidation and/or rapid thermal annealing on polycrystalline-SiGe (poly-SiGe) deposited with a LPCVD (low pressure chemical vapor deposition) system. This thermal oxidation method can directly result in the top-control oxide layer via the oxidation of amorphous-Si film and the formation of Ge nanocrystals from the poly-SiGe film. Otherwise, the rapid thermal annealing method can be also used to form Ge nanocrystals as comparison.

EXPERIMENTAL DETAILS

The Ge nanocrystal capacitors were fabricated using with two different oxidation methods.

(a). Ge Nanocrystal formation by Using One Step Thermal Oxidation

At first, a tunneling oxide of 3nm was grown on a (100)-oriented p-type Si substrate via thermal oxidation at 950^0C in a dry O_2 ambient. Then, 1nm thickness amorphous Si (a-Si) and 5nm thickness poly-SiGe film were deposited on the tunneling oxide by the LPCVD system at 475^0C under the pressure of 100 mtorr. The gas flows of GeH$_4$ and Si$_2$H$_6$ are 5 sccm, and 40 sccm, respectively. After that, a 10 nm thickness of amorphous-Si film was deposited on the sample by the LPCVD system in a Si$_2$H$_6$ gas flow rate of 40 sccm, and then performed at 900^0C thermal oxidation in a dry O_2 ambient to grow about 20 nm thick oxide layer. This one-step oxidation resulted in the top-control oxide layer via the oxidation of amorphous-Si (α-Si) film and the formation of Ge nanocrystals from the deposited poly-SiGe film. Finally, a 500 nm thickness of Al metal film was deposited and patterned to form the capacitor structures followed by a hydrogen sintering process at 400^0C for 30 min.

(b). Ge Nanocrystal Formation by Using Rapid Thermal Annealing

On the other hand, rapid thermal annealing was applied on the triple -layer structure memory capacitor. A tunneling oxide of 3nm was grown on a (100)-oriented p-type Si substrate. Then, 1nm thickness amorphous Si (a-Si) and 5nm thickness poly-SiGe film were deposited on the tunneling oxide by the LPCVD system at 475^0C under the pressure of 100 mtorr. Subsequently, a 10nm thickness CVD oxide was deposited by PECVD system and then performed with/without RTA at 900^0C for 60 sec in N_2 atmosphere as comparison. The RTA annealing can result in the formation of Ge nanocrystals from the poly-SiGe film.

RESULTS AND DISCUSSION

(a). Ge Nanocrystal formation by Using One Step Thermal Oxidation

The cross-sectional HRTEM micrograph of the top control oxide/Poly-Ge/ bottom tunneling oxide tri-layer structure for the Ge-nanocrystal sample with thermal oxidation is shown in Fig. 1. It is seen that Ge nanocrystal is wrapped within tunneling and control oxide, and Ge nanocrystal is patially continuous due to non-uniform Ge concentration distribution by long thermal oxidation time with more difficult process control. Since SiO$_2$ has higher Gibbs free energy (ΔG), 204.75 kcal/mol at 298.25 K than GeO$_2$ (-111.8 kcal/mol) [4], this indicates Ge is not likely react with O_2 than Si. Ge nanocrystal is formed and precipitated during this high temperature oxidation ambient and resulted in the condensed Ge between the two oxide layers.

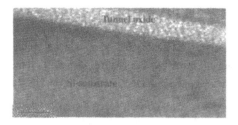

Fig 1. The cross-sectional HRTEM micrograph of the top control oxide/Poly-Ge/ bottom tunneling oxide tri-layer structure for the Ge-nanocrystal sample with thermal oxidation.

Fig. 2 shows the capacitance–voltage (C–V) hysteresis for the Ge- nanocrystal sample with thermal oxidation. At first, the hysteresis window is found with clockwise sweeping from -5 V to 5 V then back from 5V to -5 V, and it shows about 0.5V flat-band shift. However, when the sweeping voltage is increased to 10 V, the hysteresis window is also increased to 4.5 V.

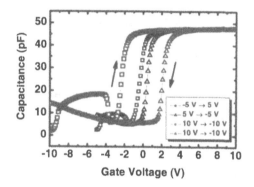

Figure 2. The capacitance–voltage (C–V) hysteresis for the sample with thermal oxidation under ±5V and ±10V counter-clockwise sweeping, respectively.

(b). Ge Nanocrystal Formation by Using Rapid Thermal Annealing

Fig. 3 shows the cross-sectional HRTEM micrograph of the top control oxide/Poly-Ge/ bottom tunneling oxide tri-layer structure for the Ge-nanocrystal sample with rapid thermal annealing. It can be seen that some Ge nanocrystals were found after rapid thermal annealing in a short time period. This is due to germanium atoms in the SiGe layer being pushed down below the top PE-oxide layer and stopped before the bottom tunnel oxide layer, during the rapid thermal annealing process with better process control. Therefore, some Ge nanocrystals can be formed and wrapped separately between the top PE-oxide layer and bottom tunnel oxide layer.

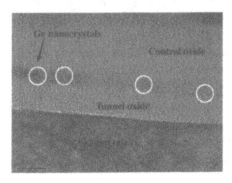

Figure 3. The cross-sectional HRTEM micrograph of the top control oxide/Poly-Ge/ bottom tunneling oxide tri-layer structure for the Ge-nanocrystal sample with rapid thermal annealing.

The capacitance–voltage (C–V) hysteresis for the sample with/without rapid thermal annealing under ±10V counter-clockwise sweeping is shown in Fig. 4. It can be seen that no hysteresis is found for the sample without RTA, but the hysteresis window is about 4 V for the sample with RTA annealing. Besides, the sample with RTA has a slightly smaller hysteresis window that with thermal oxidation.

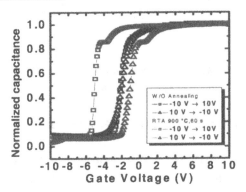

Figure 4. The capacitance–voltage (C–V) hysteresis for the sample with/without rapid thermal annealing under ±10V counter-clockwise sweeping.

Otherwise, the charge trapping density for the two kinds of samples are calculated from the following equation: $N_{charge} = (\Delta Vt) * C_{ox}$

Where ΔVt is the flat band voltage shift, and C_{ox} is the total oxide capacitance per unit area ($C_{ox}=\varepsilon_{ox}/T_{ox}$). For the one-step oxidation method, setting the total oxide thickness ($T_{ox:\ thermal}$) = 32 nm, threshold voltage shift (ΔVt) = 5V, and ε_{ox} = 3.9x8.85x10^{-14} F/cm. The charge trapping density of the Ge crystalline layer by thermal oxidation is estimated about 3.37x10^{12}cm^{-2}. For the RTA method, setting the total oxide thickness ($T_{ox:\ RTA}$) = 22 nm, and threshold voltage shift (ΔVt) = 4 V, and the charge trapping density of the Ge nanocrystals sample by RTA is estimated about 3.92x10^{12}cm^{-2}. Therefore, it can be found that although the Ge-nanocrystal sample by RTA has a slightly smaller hysteresis window (~4V shift) than that of the Ge crystalline sample by thermal oxidation (~5V shift), but the Ge nanocrystals sample by RTA with thinner oxide has a larger charge trapping ability to capture and storage charges than that by thermal oxidation.

CONCLUSION

In the paper, two oxidation methods are used to from Ge nanocrystals for flash memory. The first thermal oxidation method can directly result in the top-control oxide layer via the oxidation of amorphous-Si film and the formation of Ge nanocrystals from the poly-SiGe film. The second rapid thermal annealing method can be also used to form Ge nanocrystals as comparison. Although the thermal oxidation can simply form Ge nanocrystal in one step oxidation, but the process control is more difficult. Otherwise, the RTA annealing process with easy process control can also obtain a similar hysteresis window. Therefore, it is expected the shape of Ge nanocrystal can be well defined by good RTA process improvements.

ACKNOWLEDGEMENTS
This work was supported by the contract of NSC 96-2221-E-182-053.

REFERENCE
[1] H. I. Hanafi, S. Tiwari, and I. Khan, IEEE Electron. Devices. 43, (1996) 1553.

[2] Ya-Chin King, Tsu-Jae King, and Chenming Hu, IEDM Tech. Dig., (1998), p. 115.

[3] Z. Liu, C. Lee, V. Narayanan, G. Pei, E. C. Kan, IEEE Electron. Devices. 49, (2002) 1606.

[4] J. H. Chen, Y. Q. Wang, W.J. Yoo, Y. C. Yeo, Ganesh Samudra, Daniel SH Chan, A. Y. Du, D. L. Kwong, IEEE Electron. Devices. 51, (2004) 1840.

59

Oxide Resistive Switching Memory

Mater. Res. Soc. Symp. Proc. Vol. 1071 © 2008 Materials Research Society 1071-F07-08

Nonvolatile resistive switching characteristics of HfO$_2$ with Cu doping

Weihua Guan[1], Shibing Long[1], Ming Liu[1], and Wei Wang[2]

[1]Lab of Nano-fabrication and Novel Devices Integrated Technology, Institute of Microelectronics, Chinese Academy of Sciences, Beijing, 100029, China, People's Republic of

[2]College of Nanoscale Science and Engineering, University at Albany, Albany, NY, 12203

ABSTRACT

In this work, resistive switching characteristics of hafnium oxide (HfO$_2$) with Cu doping prepared by electron beam evaporation are investigated for nonvolatile memory applications. The top metal electrode/ hafnium oxide doped with Cu/n$^+$ Si structure shows two distinct resistance states (high-resistance and low-resistance) in DC sweep mode. By applying a proper bias, resistance switching from one state to the other state can be achieved. Though the ratio of high/low resistance is less than an order, the switching behavior is very stable and uniform with nearly 100% device yield. No data loss is found upon continuous readout for more than 10^4 s. The role of the intentionally introduced Cu impurities in the resistive switching behavior is investigated. HfO$_2$ films with Cu doping are promising to be used in the nonvolatile resistive switching memory devices.

INTRODUCTION

Recently, reversible and reproducible resistive switching phenomena induced by external electric field have been extensively studied due to its potential applications in resistive random access memories (RRAM) [1]-[14]. The typical cell of this kind of memory is a capacitor-like structure: a functional material sandwiched between two conductive electrodes. This type of memory devices can be characterized by two distinct resistance states: OFF state (with high resistance) and ON state (with low resistance). RRAM offers the possibility of high density integration, low power operation, and multilevel storage. The current candidate materials for this type of memories include ferromagnetic material such as Pr$_{1-x}$Ca$_x$MnO$_3$ (PCMO) [1], doped perovskite oxide such as SrZrO$_3$ [2] and SrTiO$_3$ [3], organic materials [4], and binary metal oxides such as NiO [5], TiO$_2$ [6], ZrO$_2$ [7], Cu$_x$O [8], and even doped SiO$_2$ [9]. Among all these candidates, binary transition metal oxides excel the others due to their simple structure, easy fabrication process and compatibility with the complementary metal-oxide semiconductor (CMOS) technology [10]. Although HfO$_2$ films are considered to be the promising gate dielectric in advanced CMOS devices, its resistive switching behavior is not fully explored, except for its unipolar switching behavior [11]-[12], i.e., turning ON and OFF occurs with the same voltage polarity. In our previous work [13], we studied the resistive switching characteristics of zirconium oxide embedded with ultrathin Au layer (with Au doping) and demonstrated that the intentionally introduced Au in ZrO$_2$ films can significantly improve the device yield. Au atoms, unfortunately, may be fatal for the integration with CMOS devices.

In this work, we report the bipolar resistive switching characteristics of HfO_2 with Cu doping for nonvolatile memory applications. Cu is nowadays emerging as an alternative to Al for metallization patterns, particularly for interconnect system having smaller dimensions and thus is friendly with CMOS technology. The top Au electrode/ HfO_2 doped with Cu/n^+ Si bottom electrode structure is fabricated and electrically characterized. The fabricated devices show excellent uniformity and stability. The possible mechanism for resistive switching is also investigated.

EXPERIMENT DETAILS

A heavily doped ($\rho \sim 3.5 \times 10^{-3}$ $\Omega \cdot$cm) n-type Si substrate is used as the bottom electrode. After chemically cleaning this n^+ silicon substrate, three sequential layers of $HfO_2/Cu/HfO_2$ (with thickness of 30/2/30 nm respectively) are deposited on the substrate via e-beam evaporation. The deposition ramp for the thin Cu layer and the HfO_2 layer is 0.2 Å/s and 0.9 Å/s, respectively. The chamber pressure is around 1.1×10^{-6} Torr. The thickness of each layer is monitored by a quartz crystal oscillator. Post-deposition annealing at 600 °C is carried out in N_2 ambient (3 L/min) to thermally diffuse the Cu atoms into HfO_2 matrix. Finally, 50 nm-thick square-shaped Au top electrodes are evaporated and defined by the lift-off process to form areas ranging from 0.36 mm^2 to 1 mm^2. For the purpose of comparison, control samples without Cu doping in HfO_2 films are simultaneously fabricated with the same process and parameter. The current-voltage (I-V) characteristics of the fabricated devices are analyzed by Keithley 4200 semiconductor characterization system in double-sweep mode, voltage list mode and sampling mode. All the measurements were performed at room temperature and under atmosphere condition. The bias voltage is applied on the top electrode with the n^+ Si bottom electrode grounded.

RESULT AND DISCUSSION

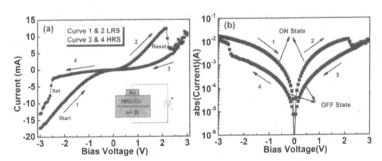

Figure 1. Typical I-V characteristics of devices with area of 0.49 mm^2 in (a) linear scale and (b) semi-log scale. The voltage is swept in the direction as follows: -3 V→0 V→3 V→0 V→-3 V, as indicated by the arrows. Inset of the upper window shows the schematic test configuration.

Figure 1 (a) shows a typical I-V curve measured by two terminal top-bottom electrodes (as illustrated in the inset of Figure 1 (a)) in linear scale. Figure 1(b) replots the same curve in semi-logarithmic scale. The voltage loop is swept as follows: $-V_{max} \rightarrow 0V \rightarrow +V_{max} \rightarrow 0V \rightarrow -V_{max}$. Hysteresis of I-V curve is clearly observed when the bias is swept back and forth, which indicates the existence of different resistance states. Here, we define low resistance state (LRS) and high resistance state (HRS) for upper and lower branches of the I–V curve, respectively. As shown in Figure 1, resistive switching from LRS (or ON state) to HRS (or OFF state) is induced by increasing the voltage up to a positive value (e.g. 2.3 V) where a decrease in current is observed. During this process, the negative differential resistance (NDR) behavior is clearly observed. The current of the OFF state increases with increasing the voltage bias in the negative direction and a switching from OFF state to ON state can be achieved at a negative bias voltage (e.g. -2.6 V). The reversible transitions between ON and OFF state can be traced for a large number of times. At a proper reading voltage (e.g. 0.3 V), the resistance of LRS and HRS can be extracted.

There are several points worthy mentioning about the I–V curve observed. First, there is no need of electroforming process for our Cu doped HfO_2 films, which is a required step by many others [5]-[7]. This is a superior property for real application. The possible reasons will be discussed in the following discussion. Second, resistive switching from LRS to HRS (turning OFF) occurs only when positive bias is applied and vice versa. This is known as the bipolar switching behavior, which means the switching is polarity dependent. This characteristic is different from the unipolar switching reported by Park et al. [11] and Lee et al. [12] with an undoped HfO_2 films. The last point is that the resistance ratio between HRS and LRS at a reading voltage (e.g. 0.3 V) is less than an order (approximately six), smaller than many other structures [2]-[6], but comparable to the BST thin films reported by Oligschlaeger et al. [14]. Despite its small signal margin for sensing different resistance states, the device yield is nearly 100%. For the total number of 1000 devices fabricated, almost every device under test shows the same reproducible switching behavior. This is consistent with the conclusion of our previous study of Au doped ZrO_2 films [13]: doping in the oxide can improve the device yield due to the more uniform and homogeneous trap concentrations. In the contrary, the control samples without Cu doping show unstable and noisy switching behaviors.

In order to evaluate the potential of our doped devices for application in nonvolatile memories, tests concerning reliability issues are performed. The repeatable resistive switching characteristic is measured in voltage list sweep mode by performing a series of consecutive write-read-erase-read (W/R/E/R) cycles. The amplitude of the write/erase voltage is ±3 V and of readout voltage is +0.3 V. As shown in Figure 2, although there is a variation of resistance values in both LRS and HRS, more than one hundred times of W/R/E/R cycles without serious sensing window deterioration are demonstrated.

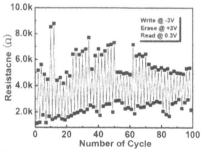

Figure 2. Endurance performance of the fabricated devices.

Figure 3 shows the capability of the devices to retain both resistance states. During this measurement, the device is first turned ON or OFF by applying -3 V or +3 V stress for a short time and then a continuous readout voltage (+0.3 V) is applied. The test scheme is briefly sketched in the inset of Figure 3. The read voltage samplings the resistance value of the device every 60 seconds. As shown in Figure 3, the resistance values of both LRS and HRS are stable and show no detectable sign of degradation over 10^4 s, confirming the nonvolatile nature and the non-destructive readout property of the devices.

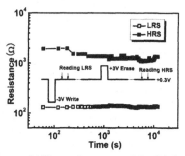

Figure 3. The nonvolatility performance of the fabricated devices.

By far, the physical origins of the resistive switching phenomena are still in debate. Hypothetical models, attempting to explain the resistive switching phenomena, can be divided into two main categories: one is the interface effect and the other is the bulk effect. As for the interface effect, Sawa *et al.* proposed a model of the alternation of Schottky barrier by carriers trapping/detrapping at the interface, causing the resistance changed [1,3]. For the bulk effect, Simmons and Verderber (SV model) proposed a model based on the conductivity modulation by charge trapping 40 years ago [15], and this model is also adopted to account for the organic memory devices [16]. Moreover, several classic conduction mechanisms in insulated materials have also been proposed, such as space charge limited current (SCLC) [8], Frenkel-Pool (F-P) emission, forming and rupture of metallic filaments (ohmic) [6] or a combination of them [2].

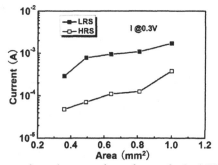

Figure 4. The current dependence on electrode area for both LRS and HRS.

According to the observed I-V characteristics of our devices, the interface effect can be firstly ruled out due to the different switching performance between doped and undoped samples, which have the same interface (fabricated under the same condition). While for the classic conduction mechanisms mentioned above, they can be distinguished through the isothermal I-V correlation [17]: linear $I\sim V$ for ohmic, $\ln I /V\sim V^{1/2}$ for F-P conduction, and $I\sim V^2$ for SCLC. According to our I-V fitting results, the ohmic and the F-P conduction can be excluded. The movement of Cu ions for the switching mechanism is not possible since the I-V correlation of ion conduction is linear. The impossibility of filament conduction (ohmic) is further confirmed by the measurement of the current as a function of the contact areas. As shown in Figure 4, the current of both LRS and HRS shows obvious area dependence, contrary to the localized filamentary conduction that has an independent current of the area. As for SCLC condution, I-V loop should be anticlockwise in the positive voltage range [8], while the I-V curve of our devices shows a clockwise behavior, as shown in Figure 1. Therefore, SCLC mechanism may not be responsible in our devices with a unique doping structure.

The experimental features of the I-V characteristics, interestingly, are identical to those described by SV model [15]: N-shaped I-V characteristic and clockwise switching in the positive voltage range. Moreover, due to the homogeneous conduction of SV model, the dependence of current on electrode area is also expected. More over, the polarity dependence of switching can be understood by SV model as follows: when electrons are injected and trapped in HfO$_2$ by applying a positive voltage (e.g., V$_{reset}$), an interior electric field is built, resulting in the decrease of the conductivity, which corresponds to the reset process. When a negative voltage (e.g., V$_{set}$) is applied, electrons are ejected out. As a result, the conductivity increases, corresponding to the set process. SV model involves carrier transportation through defects, which act as charge trapping sites. SV's description of these trapping sites is deep-level Au atoms provided by the electroforming process, which moves metal atoms from the electrode into the functional layer where they form an impurity band of charge transport levels [15]. Since we deliberately diffuse the Cu atoms into HfO$_2$ matrix during the fabrication process, the electroforming process can be exempted, as observed in the experiment. Similar phenomena of the initial low resistance state and forming-free property are also found in other doped materials [2]-[3], [18]-[19]. Due to the excellent phenomenal fitting and self-consistent explanation with SV model, we infer that the dominant mechanism for our devices may be the conductivity modulation by charge transportation in impurity band provided by Cu doping.

CONCLUSIONS

In summary, the top Au electrode/HfO$_2$ with copper doping/n$^+$ Si sandwich structure is fabricated and investigated for the nonvolatile memory applications. The intentionally introduced Cu impurities, acting as the electron traps, not only provide an effective way to improve the device yield, but also exempt the necessary of electroforming process. All the electrical observations fit the perditions of SV model quite well. The fabrication process is simple and compatible with current CMOS technology. The fabricated devices possess the properties of reversible and reproducible resistive switching, nondestructive readout, good cycling performance and nonvolatility. These excellent reliability performances indicate that Cu doped HfO$_2$ film is highly promising in the application for future nonvolatile resistive switching memory device.

ACKNOWLEDGMENTS

This work was supported by the Hi-Tech Research and Development Program of China (863 Program) under Grant No 2008AA031403, the National Basic Research Program of China (973 Program) under Grant No 2006CB302706 and the National Natural Science Foundation of China under Grant Nos 90607022 and 60506005.

REFERENCES

1. A. Sawa, T. Fujii, M. Kawasaki, and Y. Tokura, Appl. Phys. Lett., **85**, 4073 (2004).
2. C.-C. Lin, B.-C. Tu, C.-C. Lin, C.-H. Lin, and T.-Y. Tseng, IEEE Electron Device Lett., **27**, 725 (2006).
3. T. Fujii, M. Kawasaki, A. Sawa, H. Akoh, Y. Kawazoe, and Y. Tokura, Appl. Phys. Lett., **86**, 012107 (2004).
4. Y. Song, Q. D. Ling, S. L. Lim, E. Y. H. Teo, Y. P. Tan, L. Li, E. T. Kang, D. S. H. Chan, and C. Zhu, IEEE Electron Device Lett., **28**, 107 (2007).
5. J.-W. Park, J.-W. Park, K. Jung, M. K. Yang, and J.-K. Lee, J. Vac. Sci. Technol. B, **24**, 2205 (2006).
6. B. J. Choi et al., J. Appl. Phys., **98**, 033715 (2005).
7. D. Lee, H. Choi, H. Sim, D. Choi, H. Hwang, M.-J. Lee, S.-A. Seo, and I. K. Yoo, IEEE Electron Device Lett., **26**, 719 (2005).
8. A. Chen, S. Haddad, Y. C. Wu, Z. Lan, T. N. Fang, and S. Kaza, Appl. Phys. Lett., **91**, 123517 (2007).
9. C. Schindler, S. C. P. Thermadam, R. Waser, and M. N. Kozicki, IEEE Trans. Electron Devices, **54**, 2762 (2007).
10. I. G. Baek et al., in *IEDM Tech. Dig.*, 587 (2004).
11. I.-S. Park, K.-R. Kim, S. Lee, and J. Ahn, Jpn. J. Appl. Phys., **46**, 2172 (2007).
12. H.-Y. Lee, P.-S. Chen, C.-C. Wang, S. Maikap, P.-J. Tzeng, C.-H. Lin, L.-S. Lee, and M.-J. Tsai, Jpn. J. Appl. Phys., **46**, 2175 (2007).
13. W. Guan, S. Long, R. Jia, and M. Liu, Appl. Phys. Lett., **91**, 062111 (2007).
14. R. Oligschlaeger, R. Waser, R. Meyer, S. Karthäuser, and R. Dittmann, Appl. Phys. Lett., **88**, 042901 (2006).
15. J. G. Simmons and R. R. Verderber, Proc. R. Soc. Lond. A, Math. Phys. Sci., **301**, 77 (1967).
16. L. D. Bozano, B. W. Kean, V. R. Deline, J. R. Salem, and J. C. Scott, Appl. Phys. Lett., **84**, 607 (2004).
17. S. M. Sze, Physics of Semiconductor Device, 2nd ed. New York: Wiley, 1981.
18. M. Villafuerte, S. P. Heluani, G. Juárez, G. Simonelli, G. Braunstein, and S. Duhalde, Appl. Phys. Lett., **90**, 052105 (2007).
19. D. Lee, D. Seong, I. Jo, F. Xiang, R. Dong, S. Oh, and H. Hwang, Appl. Phys. Lett., **90**, 122104 (2007).

Mater. Res. Soc. Symp. Proc. Vol. 1071 © 2008 Materials Research Society　　　　1071-F08-08

Temperature Dependence of Electrical Properties of NiO Thin Films for Resistive Random Access Memory

Ryota Suzuki[1], Jun Suda[1], and Tsunenobu Kimoto[1,2]
[1]Department of Electronic Science and Engineering, Kyoto University, Kyotodaigaku-katsura, Nisikyo, Kyoto, 615-8510, Japan
[2]Photonics and Electronics Science and Engineering Center (PESEC), Kyoto University, Kyotodaigaku-katsura, Nishikyo, Kyoto, 615-8510, Japan

ABSTRACT

Temperature dependence of electrical properties in NiO thin films for ReRAM applications has been investigated. I-V measurements have been carried out in the temperature range from 100K to 523K. The resistance in the high resistance state (HRS) is almost independent of temperature below 250K, whereas it decreases with an activation energy of 300 meV above 250K. Hopping conduction and band conduction may be dominant in the low- and high-temperature range, respectively. Admittance spectroscopy on the NiO/n^+-Si structure reveals the existence of a high density of traps, which may contribute to the conduction in HRS. In the low resistance state (LRS), however, the resistance slightly increased in the whole temperature range and the trend is similar to that of metallic Ni film, indicating the metallic Ni defects is related to the conduction in LRS. The Pt/NiO/Pt structure demonstrated stable resistance switching even at temperature as high as 250°C or higher. Since other competitive nonvolatile memories will face severe difficulty in high-temperature operation, the present ReRAM shows promise for high-temperature application.

INTRODUCTION

In recent years, the resistance switching behavior of binary transition metal oxides, such as NiO [1], TiO_2 [2], has attracted much attention, owing to their potential for new-generation nonvolatile memory, Resistive Random Access Memory (ReRAM). ReRAM has advantages of miniaturization and low-power operation in comparison with other nonvolatile memories. ReRAM has another advantage of its compatibility with conventional complementary-metal-oxide-semiconductor (CMOS) technologies. However, before pushing ReRAM to industrial application, one must resolve a number of issues, including designing resistance switching characteristics and device structure. Toward this goal, it is essential to elucidate the resistance switching mechanism of the material.

Since the early 1960s, the resistance switching behavior of various transition metal oxides has been investigated [3]. These materials have two resistance states of low resistance state (LRS) and high resistance state (HRS) by applying external voltages. To explain the resistance switching phenomena, various models, such as trap charging/discharging, thermally-induced chemical reaction and Mott transition model have been proposed [2, 4-5]. However, details of the physical origins of resistance switching behavior have not been clarified yet.

It is important to establish the electrical conduction mechanism of resistance switching materials in order to find a clue for the resistance switching mechanism. Although it is useful to investigate temperature dependence of current-voltage (I-V) characteristics of resistance switching materials for understanding the conduction mechanism, there are few studies of electrical properties in the wide temperature range, especially at high temperature.

In this paper, we report electrical properties of NiO thin films prepared by a RF sputtering method. I-V characteristics in HRS and LRS in the temperature range from 100K to 523K and the conduction mechanism in LRS and HRS of NiO thin films is discussed. Moreover, we present possibility of high-temperature applications of ReRAM.

EXPERIMENT

Samples with Pt/NiO/Pt stack structure were fabricated on p-Si substrates. NiO thin films were deposited by a reactive RF sputtering method using a Ni target of 99.9% purity in Ar and O_2 gases. During sputtering, substrate temperature and working pressure were kept at 300°C and 1.5 Pa, respectively. The ratio of O_2 flow rate in the Ar + O_2 gas mixture was 5%. The thickness of NiO films was about 200 nm. From the X-ray diffraction (XRD), the deposited films have a polycrystalline structure. The chemical composition of the NiO film was determined to be $NiO_{1.07}$ by using Rutherford backscattering (RBS). After the NiO film deposition, Pt top electrodes with a 300 μm diameter were formed by electron-beam evaporation through a metal mask. I-V measurements were performed using the Keithley 4200 Semiconductor Parameter Analyzer. During I-V measurements, the bottom electrode was grounded and the bias voltage was applied to the top electrode. The temperature dependence of I-V characteristics was carried out in the temperature range from 100K to 523K.

Admittance spectroscopy was conducted in order to investigate localized states in NiO. For this measurement, an n^+-p structure, Pt/NiO/n^+-Si/Al, in which NiO is a metal deficient p-type semiconductor [6], was fabricated. The frequency dependence of conductance and capacitance were measured in the temperature range from 295K to 350K.

RESULTS & DISCUSSION

Resistance switching property of NiO thin films

Figure 1 shows the resistance switching characteristics of an NiO thin film at room temperature (RT). After forming process, which changed the initial insulating state into the low resistance state, at ~ 6.5 V, reproducible resistance switching characteristics were observed. Afterwards, by sweeping the voltage from 0 V to a positive value, an abrupt increase in the current occurred above a certain threshold voltage of 2.5 V (V_{set}). While sweeping again, a sudden drop in the current occurred at a lower voltage of 1 V (V_{reset}) than V_{set}. Here the resistance switching from HRS to LRS is called the "SET" process, and that from LRS to HRS is the "RESET" process. During the SET process, the current compliance was restricted to 10 mA to prevent the samples from break-down. The resistance switching behavior occurs regardless of the polarity of applying voltage. A typical resistance value at 0.3 V in LRS (R_{LRS}) and HRS (R_{HRS}) were about 40 Ω and 200 kΩ, respectively. The R_{HRS}/ R_{LRS} ratio was about 5000 at RT.

Figure 2 shows the logarithmic plot of I-V characteristics in Fig. 1. In LRS, the I-V characteristic was linear up to V_{reset}, which indicates that LRS is dominated by ohmic conduction. On the other hand, in HRS, I-V characteristic was linear in the low voltage region ($V < 1$ V) but slightly non-linear in the high voltage region ($V > 1$ V). In HRS, the I-V characteristic can be fitted by the following equation [7]

$$I = \frac{V}{R_0}\left\{1 + \left(\frac{V}{V_0}\right)^2\right\},$$ (1)

where the R_0 and V_0 are constants. The first term represents ohmic conduction; the second, space-charge-limited current (SCLC). It is noted that the SCLC component, which is proportional to V^3, is derived from the assumption that traps are present in NiO. Figure 2 shows the fitting results of I-V data in HRS according to the equation (1). This result indicates that the traps in NiO contribute to the conduction in HRS. At the low electric field, the injected carrier density is lower than the density of thermally generated carriers and the hopping conduction with traps may be dominant. SCLC may be dominant when the carrier injection from electrodes increases at the high electric field.

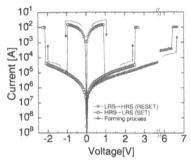

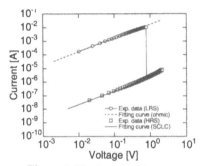

Figure 1. Resistance switching characteristics of Pt/NiO/Pt structure. The current compliance was restricted to 10 mA.

Figure 2. The logarithmic plot of I-V characteristic in Figure 1. Their fitting results using the ohmic and SCLC with a trap model are also shown.

Temperature dependence of I-V characteristics

Temperature dependence of I-V characteristics of Pt/NiO/Pt structure was investigated in the temperature range from 100K to 523K. Before the measurements, the reproducible resistance switching property was observed at RT. Then, different electrodes in a same sample were set to LRS and HRS. The measurements were performed below 0.4 V to avoid changing the resistance state. Figure 3 shows the Arrhenius plots of resistance vs. reciprocal temperature in HRS, LRS, and a Ni thin film. Temperature dependence of I-V characteristic of an Ni thin film deposited by a sputtering method was measured for comparison with that of LRS. In HRS, while the resistance was almost independent of temperature below 250K, the resistance decreased rapidly above 250K. On the other hand, the LRS resistance was slightly increased at elevated temperature.

In HRS, the activation energy of resistance (ΔE_a) estimated from the slope of Arrhenius plot is 300 meV above 250K and is 5 meV below 250K, respectively. The activation energy of 300 meV above 250K may reflect the temperature dependence of the carrier density or mobility in NiO. In the latter case, however, it is unlikely that the mobility shows the reduction by several orders of magnitude above RT. It is more reasonable that traps (or acceptors) with an ionization energy of 300 meV exist in the NiO band gap by assuming that the activation energy above 250K is due to the variation of the carrier density. In the high temperature range (> 250K), band conduction with holes thermally excited may be dominant, whereas hopping conduction through trapped holes may be dominant in the low temperature range (< 250K).

On the other hand, the LRS resistance was almost independent of temperature or slightly increased in the whole temperature range. This temperature dependence of the LRS resistance is very similar to that of a Ni thin film. The resistance switching in NiO thin films has been reported to be related to the formation and rupture of conducting filaments [8-9]. This temperature dependence of the LRS resistance indicates that conducting filaments related metallic Ni defects may exist in the NiO thin film.

We also examined the resistance switching characteristics in the temperature range from RT to 523K. The Pt/NiO/Pt structure demonstrated stable resistance switching phenomena in the entire temperature range. The HRS resistance decreased with temperature, exactly as measured R_{HRS} and R_{LRS} separately, as shown in Fig. 3, and the LRS resistance was almost independent of temperature. Figure 4 shows the resistance switching characteristics at RT and at 250°C. Although the R_{HRS}/ R_{LRS} ratio decreases with temperature, the ratio is still about 70 at 250°C, which is enough for the read dynamic margin for practical use. The present result demonstrates a promise of NiO-based ReRAM for high-temperature applications.

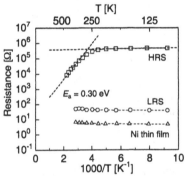 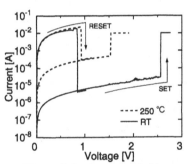

Figure 3. Arrhenius plots of resistance vs. reciprocal temperature in HRS (square), LRS (circle), and a Ni thin film (triangle). The resistance is measured at 0.3 V.

Figure 4. Resistance switching characteristics of Pt/NiO/Pt structure at RT and at 250°C.

Defect characterization in NiO by admittance spectroscopy of a NiO/n⁺-Si heterojunction

To confirm the existence of traps contributing to the conduction in HRS, admittance spectroscopy was performed in the temperature range from 295K to 350K. The conductance G and the capacitance C are given by the following equation [10]:

$$G = \frac{\omega^2 \tau}{1 + \omega^2 \tau^2} \Delta C , \qquad (2)$$

72

$$C = C_{HF} + \frac{\Delta C}{1 + \omega^2 \tau^2},$$ (3)

with

$$\Delta C = C_{LF} - C_{HF},$$ (4)

where G is the conductance of the investigated samples, G_{dc} is the DC component of the conductance, and ω is the probe angular frequency. C_{LF} and C_{HF} are the low- and high-frequency capacitances, respectively. The time constant τ is given by

$$\tau(T) = \frac{1}{N_v(T)\langle v_{th}(T)\rangle_p \sigma_p(T)} \exp\left(-\frac{\Delta E}{kT}\right).$$ (5)

Here N_v is the effective density of states in the valence band, $\langle v_{th}\rangle$ is the mean thermal velocity, σ_p is the hole capture cross section, ΔE is the activation energy of the trap, respectively.

Figure 5 shows the plot of $(G-G_{dc})/\omega$ and C measured at zero bias for the p-NiO/n$^+$-Si heterojunction as a function of frequency. The measured temperature is 295K. The $(G-G_{dc})/\omega$ and C fitted by using equations (2) and (3) are also plotted in Fig. 5. The experimental results of $(G-G_{dc})/\omega$ and C are indeed consistent with the calculated ones except for the high-frequency region. As predicted by equations (2) and (3), $(G-G_{dc})/\omega$ shows a peak and C decreases when $\omega\tau = 1$. ΔC reflects the ratio of N_t to N_s. Here N_t is the concentration of trap levels, N_s the concentration of acceptors. From the frequency dependence of $(G-G_{dc})/\omega$ and C at each temperature, N_t/N_s can be estimated to be about 7. Since the acceptor concentration N_s of NiO was determined to be 5×10^{18} cm^{-3} from C-V measurements of NiO/n$^+$-Si structure, the concentration of traps N_t can be estimated to be 4×10^{19} cm^{-3}, indicating that most holes in NiO may be trapped.

The Arrhenius plot of τT^2 vs. reciprocal temperature yields the activation energy ΔE (Figure 6). Here τ is multiplied by T^2, taking the temperature dependence of N_v and $\langle v_{th}\rangle$ into account. Then, the energy level of defects is estimated to be $E_v + 46$ meV. This value is smaller than the activation energy of 300 meV obtained by the temperature dependence of the HRS resistance in the high temperature region. The reason for this difference of the activation energy has not been clear. A deeper level may exist in NiO although it was not detected under this measuring condition.

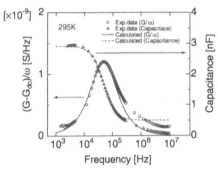 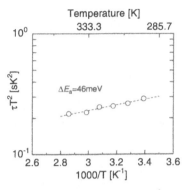

Figure 5. Frequency dependence of $(G-G_{dc})/\omega$ and capacitance of NiO/n+-Si structure measured at 295K.

Figure 6. Arrhenius plot of τT^2 vs. reciprocal temperature obtained from admittance spectroscopy.

CONCLUSIONS

The resistance switching phenomena in NiO thin films have been investigated. The current conduction mechanism has been discussed based on I-V characteristics and the temperature dependence. In LRS, metallic Ni-like filaments in NiO may contribute to the conduction. In HRS, hopping conduction and band conduction co-exist due to a high density of traps. The existence of traps has been confirmed by admittance spectroscopy of NiO/n+-Si structure. Moreover, it is experimentally demonstrated that NiO-based ReRAM can operate at 250°C or even higher. For further work, it is necessary to examine the endurance and retention properties at high temperature.

REFERENCES

1. S. Seo, M. J. Lee, D. H. Seo, E. J. Jeoung, D.-S. Suh, Y. S. Joung, I. K. Yoo, I. R. Hwang, S. H. Kim, I. S. Byun, J.-S. Kim, J. S. Choi, and B. H. Park, *Appl. Phys. Lett.*,**85**, 5655 (2004).
2. B. J. Choi, D. S. Jeong, S. K. Kim, C. Rohde, S. Choi, J. H. Oh, H. J. Kim, C. S. Hwang, K. Szot, R. Waser, B. Reichenberg, and S. Tiedke, *J. Appl. Phys.*, **98**, 033715 (2005).
3. J. F. Gibbons and W. E. Beadle, *Solid State Electron.*, 7, 785 (1964).
4. Y. Sato, K. Kinoshita, M. Aoki and Y. Sugiyama, *Appl. Phys. Lett.* 90, 033503 (2007).
5. M. J. Rozenberg, I. H. Inoue and M. J. S'anchez, *Appl. Phys. Lett.* 88, 033510 (2006).
6. H. J. Van Daal and A. J. Bosman, *Phys. Rev.*, **158**, 736 (1967).
7. R. Fors, S. I. Khartsev and A. M. Grishin, *Phys. Rev. B*, **71**, 045305 (2005).
8. D. C. Kim, S. Seo, S. E. Ahn, D.-S. Suh, M. J. Lee, B.-H. Park, I. K. Yoo, I. G. Baek, H.-J. Kim, E. K. Yim, J. E. Lee, S. O. Park, H. S. Kim, U.-I. Chung, J. T. Moon and B. I. Ryu, *Appl. Phys. Lett.,* **88**, 202102 (2006).
9. K. Jung, H. Seo, Y. Kim, H. Im, J. Hong, J.-W. Park and J.-K. Lee, *Appl. Phys. Lett.* 90, 052104 (2007).
10. J. L. Pautrat, B. Katircioglu, N. Magnea, D. Bensahel, J. C. Pfister and L. Revoil, *Solid State Electron.*, 23, 1159 (1980).

Mater. Res. Soc. Symp. Proc. Vol. 1071 © 2008 Materials Research Society   1071-F05-19

Substantial Reduction of Reset Current in CoO RRAM with Ta Bottom Electrode

Hisashi Shima[1], Fumiyoshi Takano[1], Yukio Tamai[2], Hidenobu Muramatsu[1], Hiroyuki Akinaga[1], Isao H Inoue[3], and Hidenori Takagi[3,4]

[1]Nanotechnology Research Institute, National Institute of Advanced Industrial Science and Technology, Tsukuba Central 2, 1-1-1 Umezono, Tsukuba, 305-8568, Japan
[2]Advanced Technology Research Laboratories, Sharp Corporation, 1 Asahi, Daimon-cho, Fukuyama, 721-8522, Japan
[3]Correlated Electron Research Center, National Institute of Advanced Industrial Science and Technology, Tsukuba Central 4, 1-1-1 Higashi, Tsukuba, 305-8562, Japan
[4]Department of Advanced Materials, University of Tokyo, Kashiwa, 277-8581, Japan

ABSTRACT

The resistance switching in Pt/Co-O/Pt and Ta/Co-O/Pt with the Co-O thickness of 50 nm has been investigated. Compared to Pt/Co-O/Pt, the reset current was more efficiently decreased in Ta/Co-O/Pt by using the load resistor in the forming process, indicating that the embedded resistance component with little parasitic capacitance effectively limits the current in the forming process. The reset process with the reset current lower than 0.15 mA was successfully demonstrated in Ta/Co-O/Pt. In addition, the high speed resistance switching by the voltage pulse with the pulse width of 20 ns was carried out, by investigating the pulse voltage height dependence of reset speed in Ta/Co-O/Pt.

INTRODUCTION

Metal/oxide/metal (MOM) stacking structures showing the resistance switching by applying voltage are attracting considerable attention because they are useful for the resistance random access memory (RRAM)[1-7]. In the operation of RRAM, three characteristic resistance switching process exist: forming, reset, and set. Forming is regarded as a breakdown process to activate RRAM with a relatively large forming voltage[8], which can be decreased with decreasing the thickness of the oxide layer. Reset is a resistance switching process from the low resistance state (LRS) to high resistance state (HRS), while set is a resistance switching process from HRS to LRS. One of the disadvantages in early RRAM was a large reset current exceeding 1 mA. In order to build RRAM into conventional CMOS circuits, the reduction of the reset current is required. The reduction of the instantaneous excessive current due to the forming and set is considered to be crucial in order to decrease further the reset current. In the present paper, we demonstrate the reduction of the reset current down to around 0.1 mA in RRAM with a Ta bottom electrode.

EXPERIMENT

The MOM stacking structures used in the present study was Pt(100)/Co-O(50)/Pt(100) and Ta(100)/Co-O(50)/Pt(100). The numbers in the hypothesis are the thickness of the each layer in the unit of nm. All the layers was made by the radio frequency (RF) magnetron sputtering and the oxide target was used for Co-O(50). The RF power, pressure during sputtering, oxygen content, and the substrate temperature were, respectively, 200 W, 0.3 Pa, 0 %, and RT for Co-O(50). Before sputtering Pt(100) top electrode (TE), the x-ray diffraction (XRD) measurements were carried out. In addition, the chemical bonding state of Co-O layer was investigated by x-ray photoelectron microscopy (XPS). For the current-voltage (*I-V*) measurements, the stacking structures were microfabricated into square shape with the area of 30 × 30 μm². Keithley 4200C semiconductor parameter analyzer was used in order to drive the external voltage. The external voltage was driven at TE and the bottom electrode (BE) was grounded. Instead of the current compliance by Keithley 4200C, the load resistor (LR) having the resistance R_S was connected in the series to RRAM before forming and set. LR was disconnected in the reset process. The instantaneous current in the forming process was monitored by using Agilent 6054A oscilloscope (OSC) terminated to 50 Ω. Schematic illustrations including RRAM, LR, and OSC are shown in Figs. 1(a) and 1(b). Figures 1(a) is the circuit without LR in the reset process, while Fig. 1(b) corresponds to the circuit with LR in the forming and set process.

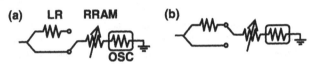

Fig. 1 Schematic illustrations for (a) the measurement system when the conventional current compliance is used and (b) the measurement system when the load resister R_S is used in order to control the transient current. RRAM and OSC correspond to resistance switching trilayer and oscilloscope, respectively.

DISCUSSION

Figures 2(a) and 2(b) show the XRD patterns for Pt(100)/Co-O(50), Ta(100)/Co-O(50). When Pt is used as BE, Pt layer shows texture with the <111> direction perpendicular to the substrate. Since distance between (111) planes in the NaCl-type CoO is close to that in Pt, a marked diffraction peaks from the (111) plane in CoO is observed in Fig. 2(a). The candidate for the diffraction peak denoted by * in Fig. 2(a) seems to be the Si (200) plane and/or CoO(111) with the space group F-43m. Note that the calculated diffraction peak from the (111) plane in CoO is close to that in Co_3O_4, which are difficult to be isolated. According to the XPS spectra, it was revealed that the Co-O layer is a mixture of CoO and Co_3O_4, rather than the single phase. The diffraction peaks of Co-O on Ta are quite different from that on Pt. Relatively broad and weak peaks from the (200) plane in CoO is observed in Fig. 2(b). The possible candidate for the diffraction peak denoted by ** in Fig. 2(b) may be CoO(111) with the space group F-43m and/or Ta(002). In our previous study, we performed the EELS measurement for the interface between Ta and CoO[9]. The core-loss EELS spectrum for O-*K* edge showed an overlap between those for Co-O and Ta oxide. In addition, the amorphous layer was observed at the interface between the Ta and Co-O layers in the TEM image. It is reasonably considered that the texture of Co-O on Ta becomes different from that on Pt because of the underlying amorphous Ta oxide layer.

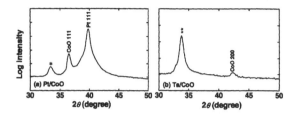

Fig. 2 XRD patterns for (a)Pt/CoO and (b)Ta/CoO.

Although crystallographic orientations of Co-O in Pt(100)/Co-O(50) and Ta(100)/Co-O(50) are different, both stacking structures of Pt(100)/Co-O(50)/Pt(100) and Ta(100)/Co-O(50)/Pt(100) exhibits reproducible resistance switching. However, Ta(100)/Co-O(50)/Pt(100) shows the different load resistor resistance dependence in terms of the reset current compared to Pt(100)/Co-O(50)/Pt(100). Figure 3(a) is I-V curves in the forming and following reset processes for Pt(100)/Co-O(50)/Pt(100). In the forming process, the load resistor of $R_S = 380$ Ω was attached. When voltage was swept up to +7.0 V, the maximum current in the forming process (I_F^{Max} denoted by the open circle in Fig. 3(a)), becomes about 17 mA. The maximum current in the following reset process (I_R^{Max} denoted by the open triangle in Fig. 3(a)) is about 20 mA. The value of I_F^{Max} is decreased by increasing R_S. However, I_R^{Max} is saturated around 1 mA for Pt(100)/CoO(50)/Pt(100) as shown in Fig. 3(b) even when $I_F^{Max} < 1$ mA. Note that several different devices were used for every value of R_S and solid symbols in Figs. 3(a) and 3(b) correspond to the averaged values. On the other hands, I_R^{Max} shows monotonic decrease with I_F^{Max} for Ta(100)/Co-O(50)/Pt(100) as shown in Fig. 3(c). Here, absolute values of I_F^{Max} were plotted in Fig. 3(c) because the negative voltage was applied in the forming process of Ta(100)/Co-O(50)/Pt(100). By using the load resistor of 68 kΩ, I_R^{Max} became less than 0.15 mA. In addition, I_R^{Max} is decreased to less than 0.1 mA for the load resistor of 150 kΩ. The possible reason for the efficient decrease of I_R^{Max} in Ta(100)/Co-O(50)/Pt(100) is the Ta oxide at the interface. Since the Ta oxide such as Ta_2O_5 is much large resistivity than Co-O, the Ta oxide layer at the Ta(100)/Co-O(50) interface would act as an embedded load resistor with little parasitic capacitance. The present results indicate that the current limitation in the forming process by the resistance with little parasitic capacitance connected in the series to RRAM is effective for decreasing the reset current in RRAM.

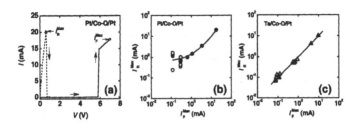

Fig. 3 (a) I-V curves in the forming and reset processes in Pt(100)/Co-O(50)/Pt(100) by the DC voltage sweep. The relation between maximum current in the forming process (I_F^{Max}) and reset process (I_R^{Max}) for (b) Pt(100)/Co-O(50)/Pt(100) and (c) Ta(100)/Co-O(50)/Pt(100).

Figure 4(a) is the I-V curves in the reset and set processes by the DC voltage sweep. Since the load resistor of 68 kΩ was connected to the corresponding forming process, the reset process with substantially reduced reset current is observed in Fig. 4(a). In the set process by the DC voltage sweep fot Ta(100)/Co-O(50)/Pt(100), the negative voltage was used because the bipolar switching was more stable in the present switching conditions compared to the nonpolar switching. This result may be related to the asymmetry of the electrode material. Note that the non-polar resistance switching is possible for Pt(100)/Co-O(50)/Pt(100), although the switching current is larger that Ta(100)/Co-O/Pt(100). In addition to the DC voltage sweep, the resistance switching by the pulse voltages was investigated. We applied various pulse voltage for the reset in Ta(100)/Co-O(50)/Pt(100). For instance, the current at 0.2 V read after applying pulse voltage with the pulse height of 1.4, 1.8, and 2.0 V was plotted as a function of the pulse width. Here, Ta(100)/Co-O(50)/Pt(100) was switched into LRS by applying the pulse voltage of -2.5V/20ns. When the pulse height was 1.4 V, Ta(100)/Co-O(50)/Pt(100) is switched into HRS with the pulse width longer than 4 ms. On the other hands, Ta(100)/Co-O(50)/Pt(100) is switched into HRS with the pulse width of 1 μs for 1.8 V, and 70 ns for 2.0 V. Shown in Fig. 4(c) is the relation between pulse voltage and pulse height required for the reset process of Ta(100)/Co-O(50)/Pt(100). In the set process, the pulse voltage of -2.5V/20ns was applied. With increasing pulse voltage height, the reset speed significantly increased. Only 1.5 V increase of the pulse voltage height leads to the 10^6 times faster reset process, indicating that the present reset process is sensitive to the voltage or power applied for the reset, being analogous to RRAM using TiON[10].

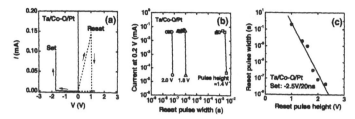

Fig. 4 (a) *I-V* curves in the reset and set processes after the forming process with the load resistor of 68 kΩ. (b) Pulse height dependence of the current at 0.2 V read following the reset pulse application for the low resistance state switched by the set pulse of -2.5V/20ns. The pulse height was 1.4, 1.8, and 2.0 V. (c) Relation between pulse width and pulse height required for the reset in Ta(100)/Co-O(50)/Pt(100).

CONCLUSIONS

The resistance switching in Pt(100)/Co-O(50)/Pt(100) and Ta(100)/Co-O(50)/Pt(100) has been investigated. By using the load resistor, the reset current was substantially decreased in Ta(100)/Co-O(50)/Pt(100) rather than Pt(100)/Co-O(50)/Pt(100). The reset process with the reset current lower than 0.15 mA was successfully demonstrated in Ta(100)/Co-O(50)/Pt(100), indicating that the embedded resistance component with little parasitic capacitance effectively limits the current in the forming process. In addition, the high speed resistance switching by the voltage pulse with the pulse width of 20 ns was carried out, by investigating the pulse voltage height dependence of reset speed in Ta(100)/Co-O(50)/Pt(100).

REFERENCES

1 S. Seo, M. J. Lee, D. H. Seo, E. J. Jeong, D.-S. Suh, Y. S. Jong, I. K. Yoo, I. R. Hwang, S. H. Kim, I. S. Byun, J.-S. Kim, J. S. Choi, and B. H. Park, Appl. Phys. Lett. **85**, 5655 (2004).

2 A. Chen, S. Haddad, Y. C. Wu, T.-N. Fang, Z. Lan, S. Avanzino, S. Pangrel, M. Buynoshki, M. Rathor, W. Cai, N. Tripsas, C. bill, M. Vanbuskirk, and M. Taguchi, *IEDM Tech. Dig.*, 746 (2005).

3 B. J. Choi, D. S. Jeong, S. K. Kim, C. Rohde, S. Choi, J. H. Oh, H. J. Kim, C. S. Hwang, K. Szot, R. Waser, B. Reichnberg, S. Tiedke, J. Appl. Phys. **98**, 033715 (2005).

4 H. Shima, F. Takano, Y. Tamai, H. Akinaga, and I. H. Inoue, Jpn. J. Appl. Phys. **46**, L57 (2007).

5 H. Shima, F. Takano, Y. Tamai, H. Akinaga, I. H. Inoue, and H. Takagi, Appl. Phys. Lett. **91**, 012901 (2007).

6 S. Q. Liu, N. J. Wu, and A. Ignatiev, Appl. Phys. Lett. **76**, 2749 (2000).

7 J. Rodrigues Contreras, H. Kohlstedt, U. Poppe, R. Waser, C. Buchal, and N. A. Pertsev, Appl. Phys. Lett. **83**, 4595 (2003).

8 K. Kinoshita, T. Tamura, M. Aoki, Y. Sugiyama, and H. Tanaka, Appl. Phys. Lett. **89**, 103509 (2006).

9 H. Shima, F. Takano, Y. Tamai, H. Muramats, H. Akinaga, I. H. Inoue, and H. Takagi, Submitted to Appl. Phys. Lett.

10 Y. Hosoi, Y. Tamai, T. Ohnishi, K. Ishihara, T. Shibuya, Y. Inoue, S. Yamazaki, T. Nakano, S. Ohnishi, N. Awaya, I. H. Inoue, H. Shima, H. Akinaga, H. Takagi, H. Akoh, and Y. Tokura, Tech. Dig. Int. Electron Devices Meet. **2006**, 793.

Mater. Res. Soc. Symp. Proc. Vol. 1071 © 2008 Materials Research Society 1071-F05-20

The Interfacial States between Metal/Oxide and the RRAM - Part II

Wei Pan, and David Russell Evans
Materials and Device Applications Lab, Sharp Labs of America, 5700 NW Pacific Rim Blvd., Camas, WA, 98607

ABSTRACT

Switching resistance of a metal/oxide/metal structure ($Pt/Pr_xCa_{1-x}MnO_3/Pt$) upon the stimulation of electric pulse has triggered vast research interests and activities for next generation resistive RAM application. Continued from an earlier paper [1] that studied the mechanism of switching resistance, this paper extends to the switching endurance discussion using admittance spectra. Experimental data indicated that there exist interfacial dipoles (or states) at metal/oxide interfaces. Switching resistance comes from the change of interfacial dipoles. However, those interfacial dipoles are all meta-stable indicating the problem of long term switching endurance. We have tested many metal/PCMO contact combinations and characterized those contacts with basic memory criteria: bit separation, data retention, switch endurance, switching speed, and readout limitations. Among them, TiN/PCMO showed large bit separation, excellent data retention, and fast switching speed, but failed long term switching endurance test. Furthermore, the poor switch endurance is discussed using energy-well diagram. Therefore, a material system providing bi-stable states with reasonably large free energy separations is a must for this type of RRAM application. These energy levels can come from either structure or electrochemical potential differences at the metal/oxide interfaces.

INTRODUCTION

Many metal/oxide/metal stacks, such as Pt/PCMO/Pt, Pt/NiO/Pt, Pt/Ti:NiO/Pt, and Pt/TiOx/Pt, have shown switching resistance between high and low states with electric pulsing [1-8]. They have been proposed for next generation non-volatile memory (RRAM) application [5-8]. It is important to understand the switching mechanism [1, 9-15] and characterize the device against non-volatile memory application criteria. In an earlier published paper [1], the authors studied Pt/PCMO contact through DC and AC electrical characterizations. The existence of an interfacial dipole layer, between Pt and PCMO, was reported. The interfacial dipole was characterized through the impedance spectroscopy plot, known as Cole-Cole plot. In combination with temperature response of Cole-Cole plots, the relaxation time constants of the dipoles were calculated and fitted very well with pulsing periods. Recently we have seen many study results attributing the resistive switching phenomenon to the metal oxide interfacial contact [9-15].

This paper, as the second part of our study, further characterizes the RRAM devices with basic memory criteria: bit separation, data retention, switch endurance, switching speed, and readout limitations. Among them, switch endurance was studied through impedance and/or admittance spectra. The interfacial dipole model [1] suggested that controlling interfacial dipole is the key to have proper memory properties. Therefore, various top metals were screened and evaluated. Results and discussions are presented here.

EXPERIMENT AND ELECTRICAL CHARACTERIZATION

Switch Endurance

Sample preparation and test procedure were described before [1]. The DC and AC characterization of Pt/PCMO/Pt film stack were carried out using a HP 4156 precision semiconductor parameter analyzer, an Agilent 81110A dual channel pulse/pattern generator, and a HP 4194A impedance/gain phase analyzer. Sample under test was first measured its original resistance (OS). Pulses, 2.8V/1μs and 5V/100ns, were then applied to reset to a low resistance state (LRS) and set to a high resistance state (HRS). The admittance spectra (100Hz to 10MHz) were measured at each resistive state: OS, LRS, and HRS.

Figure 1 shows the impedance spectra of a sample's three states. Those semi-circles inclined with an angle απ/2 respected to their resistance R axis are known as Cole-Cole plots [16, 17]. The reactance X, or the imaginary component, rises from a low negative value to a negative apex and then falls back down to a low negative value again. This apex is called relaxation point where the product of angular frequency (ω) and relaxation time constant (τ) is unit: ωτ = 1. The relaxation frequencies for all three states were around 10-100kHz indicating the dipole relaxation mechanism according to the Clausius-Mossotti equation [16, 17]. Furthermore, the dielectric loss factor, tanδ = -R/X, and dielectric constants ε', ε" can be extracted form Cole-Cole plots also [16, 17], as seen in Figures 2 and 3. Both tanδ and ε" have shown low frequency rise characters, a signature indication of interfacial dipoles relaxation [16-18]. Therefore, the behavior of those dipoles upon electrical pulses was the focus of our endurance study.

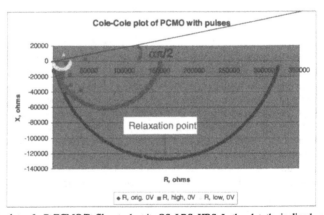

Fig. 1. Cole-Cole plots of a Pt/PCMO/Pt film stack at its OS, LRS, HRS. In the plot, the inclined angle is απ/2 where α indicate the distribution of relaxation time constant.

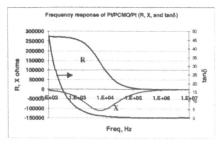

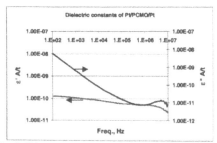

Fig. 2. The resistance, reactance, and dielectric loss factor spectra of Pt/PCMO/Pt sample.

Fig. 3. Calculated dielectric constants from admittance spectra, where A and t are the area and thickness of the sample.

Use a simple dipole model described elsewhere [16-18], the impedance Z (=R+iX) and admittance Y (=G + iB) are expressed as:

$$Z = \frac{Z_s - Z_\infty}{1 + (i\omega\tau)^{1-\alpha}} \tag{1}$$

$$Y = \frac{1}{Z} = \frac{1}{Z_s - Z_\infty} + \frac{(i\omega\tau)^{1-\alpha}}{Z_s - Z_\infty} = G_v + i\omega C_n = G + iB \tag{2}$$

where

$$G_v = \frac{1}{Z_s - Z_\infty}, \quad C_n = \frac{(i\omega)^{-\alpha}\tau^{1-\alpha}}{Z_s - Z_\infty}, \quad Z_s \text{ and } Z_\infty \text{ are the low and high frequncy values}$$

Here, α is the inclined angle $\alpha\pi/2$ respected to R axis, as shown in Fig. 1. It represents the distribution of relaxation time constants of all dipoles. Gv is a frequency independent conductance term for all physical processes that give rise to steady-state transport of charge from one electrode to the other. And Cn is a dispersive frequency dependent capacitance.

The logG-logB plot is shown in Figure 4. As marked on the plot, Gv and α could be extracted to describe two important characters of the interfacial dipoles: numbers and local configuration. Taking admittance spectra of the original, low, and high states of the sample, one could easily see how and what changes upon electrical pulses. Figure 5 describes endurance evaluation processes that use α and Gv after each set-reset cycle as the comparison quantities. Figure 5(a) is the resistance switching chart when set-reset pulses were applied. Figure 5(b) is the logG-logB plot after 1st set-reset cycle. When the sample was reset to LRS, both Gv and α were changed dramatically. This means the number of dipoles as well as local configurations was changed by the reset pulse. However, Gv and α did not go back to their original values after the set pulse. The deviation gets even bigger for the 5th and 10th switch cycles as seen in Figures 5(c) and 5(d). This is because the interfacial dipoles are meta-stable and their transition mode is uncontrollable upon electrical pulses.

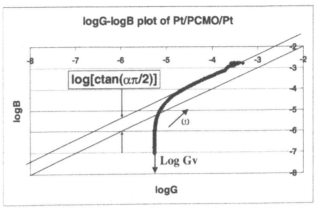

Fig. 4. A logG-logB plot of a Pt/PCMO/Pt sample. The diagonal line is logG=logB. When ω increases the plot varies from logGv to approach a line that is parallel to logG=logB line with a up shift equals log[ctg(απ/2)]. Therefore, α and Gv can be extracted from the plot to characterize ID change upon pulses.

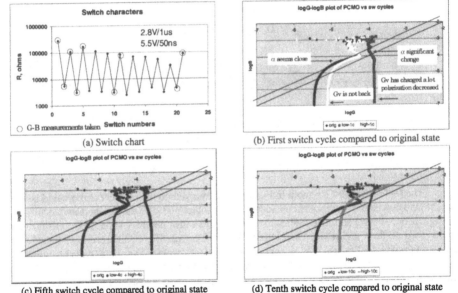

(a) Switch chart

(b) First switch cycle compared to original state

(c) Fifth switch cycle compared to original state

(d) Tenth switch cycle compared to original state

Fig. 5. The above charts describe the endurance evaluation process flow. After etch set-reset cycle, both α and Gv were changed as compared with its original state. This non-returnable phenomenon comes from uncontrollable ID transition mode. It is clear that Pt/PCMO/Pt samples don't have extended switch endurance.

Electrode Material Selection

We have tested many metals, such as Ir, Ag, Ru, Os, Pd and TiN, for PCMO hoping that one of them would form stable interfacial dipoles and controllable transition behavior. The contact between TiN/PCMO showed best switching characteristics post forming: fast switch speed, large bit separation, excellent data retention, and high yield. However, no single device passed 2000 switch cycle test. Figure 6 illustrates the test results on the TiN/PCMO/Pt devices.

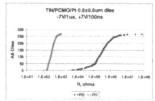 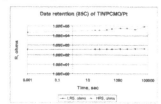 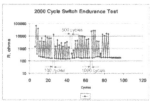

Fig. 6. TiN/PCMO/Pt stack has shown positive results post forming in terms of bit separation, data retention, but did not pass 2000 cycle switch endurance test. The device size was 0.6x0.6 μm.

DISCUSSIONS

Interface defects are believed to be the source of the interfacial dipoles [18]. They respond to an external electrical pulse in the following way. If the pulse period fits a dipole's relaxation time constant and the pulse voltage is greater than a threshold value, an ionic jump is then occurred causing dipole transition from one state to another. The interface remains electrically neutral before and after the transition. This neutrality is important for data retention. Charged interface, caused by electrons trapping for instance, will generate large internal field causing loss of electrons from those traps. External disturbs, such as read operation, will eventually cause data loss. Therefore, from data retention point of view, two energy levels with different atomic configurations at the interface are indispensable for RRAM devices. Oxygen vacancies were mentioned to be one of the common interfacial defects that would jump across metal/oxide boundary [12-15]. The displacement of vacancies across boundary changes chemical potential, which would induce electrical potential change according to Enerst equations (superscripts e and o indicates electrode and oxide, A represents vacancies):

$$\phi^{(e)} - \phi^{(o)} = -\frac{1}{F}(\mu_{A^+}^{(e)} - \mu_{A^+}^{(o)}) \qquad (3)$$

$$\mu_{A^+} = \mu_{A^+}^0 + RT \ln \gamma C_{A^+} \qquad (4)$$

Using the experimental data of reducing conduction current with DC bias, shown in Figure 7, we can calculate the barrier height change across boundary due to vacancy displacement, which is also known as electrode polarization [16-18].

Aforementioned two energy level structure could be sketched conceptually in Figure 8, which is quite different from currently known types of non-volatile memories: FeRAM, MRAM, and Flash memory. For FeRAM and MRAM, bits are written in dual but equal energy wells. The position and level of those energy wells are strictly defined by crystal fields: switch is fast, bit separation is large, data retention is good, and endurance is excellent. The flash memory, on the other hand, is a single well system. Two non-volatile states are identified between empty well or electron filled well. Here electrical isolation to prevent electron loss is critical for data retention.

Characteristics of RRAM are different because of the dual-unequal energy wells. First, the switch speed of HRS-LRS is different from LRS-HRS due to different activation energies ΔE_{a1} and ΔE_{a2}. This implies a switching endurance problem. Second, the energy levels are meta-stable resulting varying bit levels and bit separation. Therefore, select a material system that has two separated but thermodynamically stable energy levels is indispensable. Oxygen vacancies move across metal/oxide boundary is known as redox process in electrochemistry. As shown in Fig. 7, the redox has induced barrier height change for about 0.1eV in our sample. In a rechargeable battery system, reduction and oxidation states of cathode and anode have well defined electrochemical potential levels. When external electrical potential exceeds a threshold level, the redox process happens, as one see in those cyclic voltammetry curves. Use redox concept in electrochemistry to design the material system may lead the way for RRAM applications.

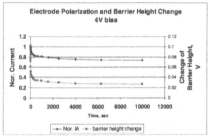

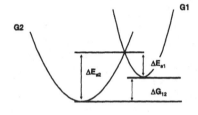

Fig. 7. The change of barrier height calculated from
$$\Delta\phi = \frac{kT}{e}\ln(1 - \frac{I}{I_0})$$ is as large as 0.076V due to displacement of vacancies (or known as electrode polarization).

Fig. 8. An illustration of dual but not equal energy well model for RRAM type of memory. Characteristics of RRAM are related to those energy levels. Bit separation is determined by ΔG_{12}; Switch speed is proportional to the activation energies, ΔE_{a1} and ΔE_{a1}.

CONCLUSIONS

In this paper, authors use Cole-Cole plot and admittance spectra to study switch endurance of Pt/PCMO/Pt devices. Results indicate that the interfacial dipoles are the source of varying resistance upon electric pluses. The dipoles are all meta-stable resulting poor switch endurance. Although TiN/PCMO has shown better RRAM performance, the switch cycles did not exceed 2000. Authors also elaborate the reasons of having poor endurance for this type of devices by using energy well concept. Authors point out that to have thermodynamically stable energy levels is the key factor for NVM. Redox process in electrochemistry could provide such stable levels at cathode/anode, which could be used to design RRAM material system.

REFERENCES

1. W. Pan and David R. Evans, Science and Technology of Nonvolatile Memories, edited by O. Auciello, J. Van Houdt, R. Carter, S. Hong (Mater. Res. Soc. Symp. Proc. 933E, Warrendale, PA, 2006), 0933-G04-05
2. Y. Watanabe, J.G. Bednorz, A. Bietsch, Ch. Gerber, D. Widmer, and Beck, Appl. Phys. Lett. 78, No.23, 3738(2001).
3. K Aoyama, K. Waku, A. Asanuma, Y. Uesu, and T. Katsufuji, Appl. Phys. Lett. 85, No. 7, 1208(2004).
4. A. Odagawa, H. Sato, I. H. Inoue, H. Akoh, M. Kawasaki, and Y. Tokura, Phys. Rev. B, 70, No. 22, 4403(2004).
5. W. W. Zhuang et. al., International Electron Devices Meeting, Technical Digest (Cat. No. 02CH37358) (IEEE, Piscataway, NJ, 2002), pp193-196.
6. I. G. Baek et. al., International Electron Devices Meeting, Technical Digest (Cat. No. 05CH37703) (IEEE, Piscataway, NJ, 2005), pp769-772
7. K. Tsunoda et. al., International Electron Devices Meeting, Technical Digest (Cat. No. 07CH37934) (IEEE, Piscataway, NJ, 2007), pp767-770
8. Myoing-Jae Lee, et. al., International Electron Devices Meeting, Technical Digest (Cat. No. 07CH37934) (IEEE, Piscataway, NJ, 2007), pp771-774
9. A. Baikalov, Y. Q. Wang. B. shen, B. Lorena, S. Tsui, Y. Y. Sun, T. T. Xue, and C. W. Chu, Appl. Phys. Lett. 83, No.5, 957(2003).
10. S. Tsui, A. Baikalov, J. Cmaidalka, Y. Y. Sun, Y. Q. Wang, Y. Y. Xue, C. W. Chu, L. Chen, and A. J. Jacobson, Appl. Phys. Lett. 85, No.2, 317(2004).
11. R. Y. Gu, Z. D. Wang, and C. S. Ting, Phys. Rev. B 67, No.15, 3101(2003).
12. A. Sawa, T.Fujii, M. Kawasaki, Y. Tokura, Appl. Phys. Lett. 88, No. 23, 2112(2006)
13. A. Sawa, T. Fujii, M. Kawasaki, and Y. Tokura, Jpn. J. Appl. Phys. 44, No. 40, 1241(2005)
14. T. Fujii, et. al., Phys. Rev. B 75, No. 16, 5101(2007)
15. S. Muraoka, et. al., International Electron Devices Meeting, Technical Digest (Cat. No. 07CH37934) (IEEE, Piscataway, NJ, 2007), pp779-782
16. L.L Hench and J.K. West, "Principles of Electronic Ceramics," John Wiley & Sons (1990) pp.144-146
17. W. D. Kingery, "Introduction to Ceramics" Second Edition, John Wiley & Sons (1976) pp.913-972
18. "Electrode Processes in Solid State Ionics," *Theory and Application to Energy Conversion and Storag,* Proceedings of the NATO Advanced Study Institute held at Ajaccio (Corsica), 28 August - 9 Sept 1975, ed. M. Kleitz and J. Dupuy, D. Reidel Publishing Company (1975), pp.149-183

Mater. Res. Soc. Symp. Proc. Vol. 1071 © 2008 Materials Research Society   1071-F09-12

Reproducible Electro-resistance Memory Effect in Ag/La0.67Sr0.33MnO3 Thin Films

Lina Huang, Bingjun Qu, and Litian Liu
Institute of Microelectronics, Tsinghua University, Beijing, 100084, China, People's Republic of

ABSTRACT

The hysteretic and reproducible electro-resistance memory effect has been investigated in epitaxial $La_{0.67}Sr_{0.33}MnO_3$ (LSMO) films under DC-bias stress and voltage pulses. The bias-sensitive current-voltage characteristic of the Ag/LSMO system is distinctly nonlinear, asymmetric and hysteretic, which indicates the appearance of the resistive switching. The pulsed voltage amplitude and duration dependence of the nonvolatile resistive switch were also provided. Clear resistance switching cycles were observed at room temperature under voltage pulses of ±5V and ~150ns. Reproducible switching properties, involving voltage-induced stepwise resistance change, resistance state saturation, and pulse duration dependent multilevel switchable capability, demonstrate well controllability with respect to future nonvolatile random access memory applications.

INTRODUCTION

In the last decade, intensive efforts have been exerted to the unique room temperature electric-pulse-induced resistance (EPIR) change effect discovered in numerous transition-metal-oxides, such as perovskite manganites [1-4], titanates [5,6], zirconates [7,8] and niobic oxides [9-11]. The resistance of the oxides can reversibly switch between two stable resistive states with applied short electric pulses of different polarity, and the modified resistance retains even after removing the pulse source. The fascinating features of the EPIR effect manifest a great potential for digital device applications, e.g. nonvolatile resistance random access memories (RRAM) with low power consumption, fast write and erase speed, simple device structure, and the possibility of easy scale-down [12].

As for the microscopic mechanism of the bi-stable resistive switching, various models have been proposed, involving field-driven lattice distortions [1], Schottky barriers with interfacial states [3], electrochemical migration at the metal-oxide interface [2,4], and phase separation, etc. However, the above models could not explain the whole experimental results of the resistive switching. It is hence of vital importance to systematically study the diverse behaviors of the resistive switch, which will provide another clue to understanding the effect. In the present study, the hysteretic and polarity-dependent resistive switch properties of Ag-$La_{0.67}Sr_{0.33}MnO_3$-$SrTiO_3$ system were investigated. The nonlinear, asymmetric and hysteretic current-voltage characteristic indicates the appearance of the resistive switching, and the reproducible switching properties share strong dependence on different selections of pulse parameters, thus demonstrating flexible controllability and multilevel storage capability as for future nonvolatile memory applications.

EXPERIMENT

200nm thick LSMO films were grown on (001) $SrTiO_3$ (STO) single crystal substrates by a pulsed laser deposition (PLD) process using a KrF excimer laser (248nm in wavelength). The base pressure for ablation was prepared below 8×10^{-6} Torr. The substrate temperature was held at 770℃ under an oxygen partial pressure of 48Pa during ablation. The repetition rate of laser pulse was 3Hz, and the laser irradiance at the target was fixed at $1.8J/cm^2$. After deposition, the film was furnace cooled to the ambient temperature at a speed of 7℃/min in one atmosphere of oxygen. Silver contact pads of 0.1mm radius with 0.25mm spaced apart were attached onto the LSMO layer using silver paint.

A schematic diagram of the device structure and the measurement circuit is shown in figure 1. The Tektronix TM 5003 pulse generator and the Keithley 2400 source meter unit were both controlled by a software program. The two-terminal resistance was examined after applying the electric pulses onto both Ag electrodes by feeding a 5×10^{-6}A direct current and reading the corresponding voltage. Current–voltage characteristics were also tested by the Keithley measurement unit. The film crystallinity was characterized by four-circle X-ray diffraction (XRD, D/MAX-RB Diffractometer), while the film morphology was detected using atomic force microscopy (AFM).

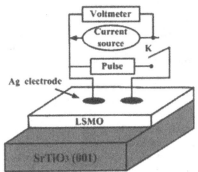

Figure 1. The schematic configuration of the Ag-La$_{0.67}$Sr$_{0.33}$MO$_3$-SrTiO$_3$ device.

RESULTS AND DISCUSSIONS

Structural and surface morphology characterization

Epitaxial growth of LSMO films was evidenced in XRD analysis. Figure 2 plots the XRD spectra of the in-situ grown LSMO film. It is obviously shown that the film is of single phase and highly c-axis oriented, with a strong peak identified as LSMO (002) besides the STO (002) diffraction peak. The LSMO (002) peak gives a full width at half maximum (HWHM) of 0.28° in the rocking curve (inset of figure 2), exhibiting good crystallinity.

The surface morphology of the LSMO films was measured by AFM, and the result is presented in figure 3. The film is very smooth with a mean square root roughness of 1.71nm and no obvious great particles or clusters are detected.

Carrier transport properties

The I –V characteristics obtained at room temperature are plotted in figure 4, with increased voltage range from ±1V to ±5V to verify the dependence of the DC bias stress. The voltage scan direction is 0V→+Vmax→0V→–Vmax→0V. It can be seen that the curve exhibits nonlinear and asymmetric behaviors along with a clear hysteresis both at positive and negative bias, reasonably consistent with the previous reports [3,13,14]. As the bias voltage increases, the nonlinear, asymmetric and hysteretic responses become more notable, suggesting the bias-sensitivity and appearance of the resistive switching. The nonlinearity and asymmetry indicate possible Schottky barriers at the Ag/LSMO interface although it would play a minor role in the resistive switching. The hysteresis captured here is quite diffusive and is likely ascribed to a large intra-system charge transfer for the electronic inhomogeneous heterostructure [15]. It should be pointed out that the virginal resistance of the sample is on the order of kΩ in magnitude, as illustrated in figure 4. Consecutive I –V sweep (>20 cycles) can gradually reduce the resistance, and stabilize it at about 330Ω finally. Such a bias stress process to get the stable resistance state with hysteretic I –V characteristics is suggested to be related to one kind of accumulation/alignment process of interfacial carriers.

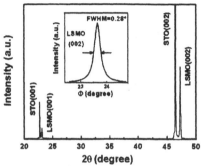

Figure 2. The X-ray diffraction pattern of in-situ grown LSMO film deposited on STO (001) substrate. The inset shows the rocking curve of the LSMO (002) reflection.

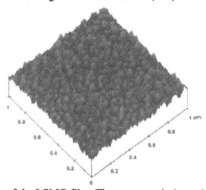

Figure 3. The AFM image of the LSMO film. The scan area is 1μm×1μm.

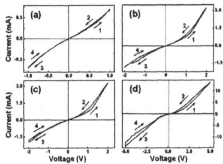

Figure 4. I –V characteristic of Ag/LSMO/STO structure. Arrows in the figure denote the voltage scan direction.

Electric-pulse-induced resistance switching behavior

Figure 5 shows the typical stepwise resistance switching result of the Ag/LSMO/STO structure driven by electric pulses. When exposed to a train of voltage pulses of ±5V/150ns, the Ag/LSMO sample starts to switch alternatively between two resistive states, and both of the states are nonvolatile at room temperature. The resistance change ratio (termed here EPIR ratio, defined as (Rmax-Rmin)/Rmin, where Rmax and Rmin are the maximum and minimum resistance induced by electric pulses, respectively) increases steeply with increasing pulsed voltage until a saturation value of about 60% is achieved at 12V (inset of figure 5). Part of the reason for the saturation is that the carriers at the metal-oxide interface inject into the interfacial microdomains to attain high resistance state, resulting in full occupation of the domains [16]. It is worth noting that different initial resistance states correspond to different stepwise figure profile, thus higher EPIR ratio and lower threshold voltage are expected as initial forming process is improved [3,17].

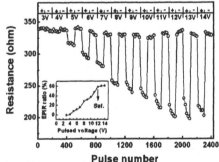

Pulse number

Figure 5. Stepwise resistance change as a function of pulsed voltage. The lower inset shows the relationship between EPIR resistance change ratio and pulsed voltage.

The pulse duration dependent multilevel switch properties are further investigated. It is confirmed from figure 6 that a given resistive state increases (or decreases) cumulatively with application of multiple negative (or positive) voltage pulses, exhibiting a stair step-up or step-down like resistance transition. Upon reversing the pulse direction, a similar step-down or step-up return to the opposite resistive state takes place. As is seen from figure 6, the sample can be set to different discrete intermediate resistance levels and each of the resistance levels has a specific state of memory. As a result, multilevel switchable states can be extracted from the Ag/LSMO/STO memory unit. It is expected that any target resistance state can be addressed at will by application of appropriate pulse number and pulse amplitude.

Compared to the 50ns pulse duration case (figure 6(a)), the stair step resistance transition driven by 5μs pulses (figure 6(b)) is quite different. The step-up transition starts at a larger threshold voltage (about -7V) and is diffusive. During the step-down process, the threshold voltage is lower (about +5V) and the transition is sharper indicated by distinct discontinuity points at different pulsed voltage. It is therefore believed that the low to high multilevel resistance switch is available under applying narrow negative pulses, while the high to low multilevel switch is easily driven by wide positive pulses.

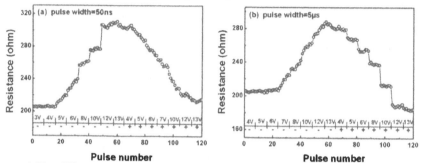

Figure 6. The different multilevel switch properties versus pulse number and pulse amplitude by application of 50ns pulses (a) and 5μs pulses (b).

In the case of Ag/LSMO/STO system, there exist diverse resistance switching behaviors modulated flexibly by pulse number, pulsed voltage amplitude and duration. Reasonably controlling the application modes and parameters of voltage pulses can not only obtain optimized switching characteristics for device applications, but also provide another way to elucidating the resistive switching mechanism.

CONCLUSIONS

The hysteretic and reproducible electro-resistance memory effect of epitaxial $La_{0.67}Sr_{0.33}MnO_3$ films deposited by a PLD process has been studied for nonvolatile memory applications. A bias-sensitive current-voltage characteristic with distinct nonlinearity, asymmetry and hysteresis was observed, which indicates appearance of the resistive switching. Under voltage pulses of ±5V amplitude and 150ns duration, clear resistance switching cycles were obtained. By different selections of electric pulses, diverse behaviors of the resistive switching were also investigated, which show a great potential for the well control of the EPIR effect in response to practical device applications. The physical details and the optimization of the resistive switch are now under further investigation.

ACKNOWLEDGMENTS

The authors are pleased to acknowledge C. Yang and G.L. Xie for giving support to measurement. This work is supported by National Natural Science Foundation (NNSF) Grant No. 90407013, and in part by the Ministry of Science and Technology of the People's Republic of China (863) Grant No. 2006AA03Z317.

REFERENCES

1. S. Q. Liu, N. J. Wu and A. Ignatiev, *Applied Physics Letters* 76 (19), 2749-2751 (2000).
2. A. Baikalov, Y. Q. Wang, B. Shen, B. Lorenz, S. Tsui, Y. Y. Sun and Y. Y. Xue, *Applied Physics Letters* 83 (5), 957-959 (2003).
3. A. Sawa, T. Fujii, M. Kawasaki and Y. Tokura, *Applied Physics Letters* 85 (18), 4073-4075 (2004).
4. S. Tsui, A. Baikalov, J. Cmaidalka, Y. Y. Sun, Y. Q. Wang, Y. Y. Xue, C. W. Chu, L. Chen and A. J. Jacobson, *Applied Physics Letters* 85 (2), 317-319 (2004).
5. Dooho Choi, Dongsoo Lee, Hyunjun Sim, Man Chang and Hyunsang Hwang, *Applied Physics Letters* 88, 082904 (2006).
6. B. J. Choi, D. S. Jeong, S. K. Kim, C. Rohde, S. Choi, J. H. Oh, H. J. Kim, C. S. Hwang, K. Szot, R. Waser, B. Reichenberg and S. Tiedke, *Journal of Applied Physics* 98, 033715 (2005).
7. Min Kyu Yang, Dal-Young Kim, Jae-Wan Park and Jeon-Kook Lee, *Journal of the Korean Physical Society* 47, S313-S316 (2005).
8. Dongsoo Lee, Hyejung Choi, Hyunjun Sim, Dooho Choi, Hyunsang Hwang, Myoung-Jae Lee, Sun-Ae Seo and I. K. Yoo, *IEEE Electron Device Letters* 10.1109/LED.2005.854397 (2005).
9. S. Seo, M. J. Lee, D. H. Seo, S. K. Choi, D. S. Suh, Y. S. Joung, I. K. Yoo, I. S. Byun, I. R. Hwang, S. H. Kim and B. H. Park, *Applied Physics Letters* 86, 093509 (2005).
10. S. Seo, M. J. Lee, D. C. Kim, S. E. Ahn, B. H. Park, Y. S. Kim, I. K. Yoo, I. S. Byun, I. R. Hwang, S. H. Kim, J. S. Kim, J. S. Choi, J. H. Lee, S. H. Jeon and S. H. Hong, *Applied Physics Letters* 87, 263507 (2005).
11. Jae-Wan Park, Jong-Wan Park, Dal-Young Kim and Jeon-Kook Lee, *Journal of Vacuum Science Technology A* 23 (5), 1309-1313 (2005).
12. W. W. Zhuang, W. Pan, B. D. Ulrich, J. J. Lee, L. Stecher, A. Burmaster, D. R. Evans, S. T. Hsu, M. Tajiri, A. Shimaoka, K. Inoue, T. Naka, N. Awaya, K. Sakiyama, Y. Wang, S. Q. Liu, N. J. Wu and A. Ignatiev, *IEEE International Electron Device Meeting*, 193-196 (2002).
13. A. Odagawa, T. Kanno, H. Adachi, H. Sato, I. H. Inoue, H. Akoh, M. Kawasaki and Y. Tokura, *Thin solid films* 486, 75-78 (2005).
14. Joe Sakai and Syozo Imai, *Journal of Applied Physics* 97, 10H709 (2005).
15. M. J. Rozenberg, I. H. Inoue, M. J. Sanchez, *Applied Physics Letters* 88, 033510 (2006).
16. Tong Lai Chen, Xiao Min Li, Rui Dong, Qun Wang, Li Dong Chen, *Thin solid films* 488, 98-102 (2005).
17. T. Fujii, M. Kawasaki, A. Sawa, H. Akoh, Y. Kawazoe and Y. Tokura, *Applied Physics Letters* 86, 012107 (2005).

Organic Resistive Switching
Memory

Mater. Res. Soc. Symp. Proc. Vol. 1071 © 2008 Materials Research Society   1071-F06-04

Resistive Electrical Switching of Cu+ and Ag+ based Metal-Organic Charge Transfer Complexes

Robert Mueller[1], Joris Billen[1,2], Aaron Katzenmeyer[1], Ludovic Goux[3], Dirk J. Wouters[3], Jan Genoe[1], and Paul Heremans[2,4]

[1]PT\SOLO\PME, IMEC v.z.w., Kapeldreef 75, Leuven, 3001, Belgium
[2]ESAT, Katholieke Universiteit Leuven, Kasteelpark Arenberg 10, Leuven, 3001, Belgium
[3]PT\CMOSRD\MEMORY, IMEC v.z.w., Kapeldreef 75, Leuven, 3001, Belgium
[4]PT\SOLO, IMEC v.z.w., Kapeldreef 75, Leuven, 3001, Belgium

ABSTRACT

Memory cells based on Cu^+ and Ag^+ metal-organic charge-transfer complexes, as for example CuTCNQ (where TCNQ denotes 7,7',8,8'-tetracyanoquinodimethane), are well known for their bistable resistive electrical switching since 1979. The switching mechanism however remained unclear for very long time. In this contribution we describe the different views (bulk vs. interfacial switching), give evidence for interfacial switching in the case of CuTCNQ, and present a model allowing explaining the bipolar resistive electrical switching by an interfacial effect, even for experiments considered until now as proof for bulk switching. The proposed switching mechanism is based on bridging of an ion-permeable layer (or gap) by conductive Cu channels, which are formed and dissolved by an electrochemical reaction implying monovalent Cu^+ cations, originating from a solid ionic conductor (as for example CuTCNQ). The model was furthermore generalized to other memory systems consisting of a permeable layer and a solid ionic conductor, including also inorganic solid ionic conductors as for example Ag_2S.

INTRODUCTION

Downscaling of traditional charge-storage based non-volatile memory technology (Flash) becomes more and more challenging due to physical limitations and increasing processing complexity. According to the International Technology Roadmap for Semiconductors [1] NOR and NAND Flash Memories will encounter scaling issues around the years 2009 and 2013, respectively (Fig. 1).

Resistive switching memories are not expected to suffer from these scaling issues and are therefore currently investigated as potential candidates for future memory applications. Various kinds of resistive switching memories have been proposed so far, as for example the oxide resistive random access memory [2], the phase change memory [3], the programmable metallization cell [4,5] and the chalcogenide based "Nanobridge®" [6]. The working principle of these different memory types have been compared in a recent paper [7].

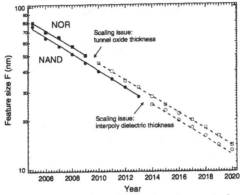

Figure 1. International Technology Roadmap for Semiconductors (redrawn from the 2006 update [1]).

Several organic materials have been investigated for resistive switching memory cells as well (see [8-10] for review). CuTCNQ (where TCNQ denotes 7,7',8,8'-tetracyanoquino-dimethane) is one of the most known organic materials used for memory applications since the Potember et al. paper in 1979 describing bistable resistive switching of CuTCNQ based memory cells with Cu bottom contacts and (evaporated) Al top contacts [11]. Recent studies demonstrated that CuTCNQ based large area memory cells can exhibit up to 10000 write/erase cycles [12,13], retain the ON state for up to 60h [13], and are operational up to 80°C [13]. Furthermore, electrical switching of CuTCNQ nanowires integrated into 250 nm diameter contact holes (vias) of Cu CMOS BEOL wafers has been reported [14] and solution based CuTCNQ nanocrystal and "mushroom" growth in even smaller vias was possible [15].

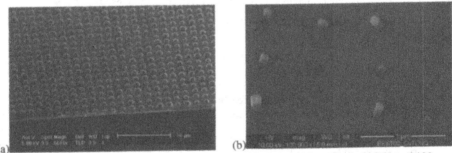

Figure 2. (a) CuTCNQ "mushrooms" and (b) nanocrystals grown respectively in 250 and 100 nm diameter contact holes [15].

Thorough investigation of scientific literature on CuTCNQ based memories however revealed that the endurance reported by many groups is significantly lower than 10000 write/erase cycles [12,13]. Furthermore, a large spread in threshold voltages, switching currents,

ON/OFF current ratio's etc. was found between devices prepared by different research groups as well as different CuTCNQ growth methods within the same group.

In addition a dependence of the threshold voltage with the CuTCNQ layer thickness would be expected for bulk switching of CuTCNQ based memory. However, this seems not be the case: threshold voltages for OFF→ON and ON→OFF switching were nearly independent upon the CuTCNQ layer thickness (50 to 300 nm) [16].

Furthermore, in contrast to most other organic memories and devices where smooth layers are expected to yield best results, we found best endurance (2500 write/erase cycles) for CuTCNQ nanowire structures with very high surface roughness [17].

All these observations, suggesting that resistive electrical switching of CuTCNQ based memories with aluminum top contacts might not be a simple bulk phenomenon, led us to perform a close investigation of the CuTCNQ switching process.

LITERATURE OVERVIEW ON CuTCNQ SWITCHING

Potember et al. interpreted the resistive electrical switching of CuTCNQ based memory cells as bulk phenomena, as for example a phase transition [11,18]. This view was afterwards confirmed by Kamitsos et al. who observed by Raman spectroscopy a decrease of the signal of the TCNQ⁻ anion and the appearance of neutral TCNQ after programming a CuTCNQ memory for several hours to the ON state [19]. From this observation the following switching mechanism (eq. 1) was proposed, implying partial transformation of ionic CuTCNQ (OFF state) into neutral Cu and neutral TCNQ (ON state) in presence of an electric field (ΔE):

$$\underset{\text{high impedance ("OFF" state)}}{[Cu^{+}TCNQ^{-}]_{n}} \quad \overset{\Delta E}{\leftrightarrow} \quad \underset{\text{low impedance ("ON" state)}}{Cu^{0}_{x} + (TCNQ^{0})_{x} + [Cu^{+}TCNQ^{-}]_{n-x}} \quad (eq.1)$$

Several years later Matsumoto et al. [20] observed electrical switching of CuTCNQ using a conductive scanning tunneling microscope tip as top contact, and Yamaguchi et al. [21] reported the growth of conductive hillocks during similar experiments. This last result [21] was also interpreted as supporting bulk switching of CuTCNQ. Later, Heintz et al. proved in 1999 that CuTCNQ can exist in 2 different crystalline phases (denoted as phase I and phase II) and suggested that the resistive electrical switching might be related to a phase transition, if the switching is not a consequence of CuTCNQ roughness [22]. Very recently Hefczyc et al. [23] reported electrical switching of CuTCNQ by using probe needles of various metals as top contacts, which might also be considered as an indication for bulk switching.

All these experiments were suggestive for bulk switching of CuTCNQ. On the other hand several papers presented evidence for interfacial switching of CuTCNQ memories. Sato et al. [24] for example performed impedance and capacitance measurements and attributed the switching process to an interfacial effect at the CuTCNQ\Al interface. This is in good agreement with Hoagland et al. observation that the resistance of Cu\CuTCNQ\Al junctions (1 mm^2)

99

increased from 0.05-15 Ω under vacuum to 10^2-10^4 Ω after air exposure [25]. Furthermore, recent impedance spectroscopy measurements performed by Kever et al. [26] also confirmed a highly resistive CuTCNQ\Al interface. In addition, Oyamada et al. [16] reported reproducible switching of ITO\Al (Al₂O₃)\CuTCNQ\Al with a deliberately added Al₂O₃ layer. Recently Hefczyc et al [23] were able to reproduce this switching for Cu\CuTCNQ\Al₂O₃\Au and similar devices in which the Al₂O₃ interlayer was replaced by ZrO₂ (stabilized by TiO₂) or Y₂O₃.

As shown in this overview (summarized in Fig. 3), several experiments on CuTCNQ switching reported in the literature are in agreement with a bulk switching while others are in favor of an interfacial effect. In the next section we will describe critical experiments from which we developed a general model for resistive electrical switching of CuTCNQ and Cu⁺ and Ag⁺ based metal-organic charge transfer complexes.

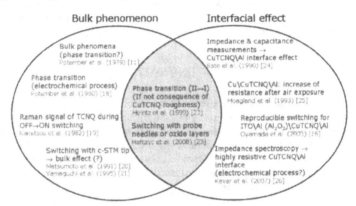

Figure 3. Literature overview: bulk versus interfacial switching

RESULTS AND DISCUSSION

Typical CuTCNQ memory cells were prepared by sandwiching the material between Cu, Au or Pt bottom contacts and Al (evaporated) top contacts. A previous study by Hoagland et al. [25] showed that for Cu\CuTCNQ\M junctions (with M=Al, Cr, and Cu) a large increase of resistance was observed after transferring the elements with Al top contacts from vacuum to air, whereas no significant increase in resistance was observed for junctions with Cr or Cu top contacts. In order to study the influence of the top contact on the switching behavior of CuTCNQ memory cells we investigated Au\CuTCNQ nanowire\M cells (prepared as described in [17]) with top contacts (M) of Al, Ti, Yb, Ni, and Au. Electrical measurements on the corresponding memory cells (Fig. 4) in air revealed that CuTCNQ typical switching (switching to the ON state by applying a negative voltage to the top contact and switching to the OFF state by inversing the polarity [13,14,17]) was observed for Al, Ti, and Yb, but not for Ni and Au.

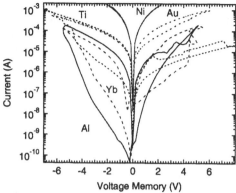

Figure 4. Log|I|-V curves of Au\CuTCNQ nanowire\M elements (measured in air) with various top contact metals. Typical bipolar resistive electrical switching was observed for M= Al, Ti and Yb, but not for Ni and Au.

Since Al, Ti and Yb are much easier oxidized than Ni and Au, and because Hoagland et al. reported an increase of resistance of Cu\CuTCNQ\Al junctions after air exposure [25], we performed additional experiments where the memory cells were directly measured under protective atmosphere after vacuum deposition of the top contacts. These experiments, performed with Cu\CuTCNQ\Al and Cu\CuTCNQ\Yb cells, indicated that the devices were in an initial highly conductive state after transfer from vacuum to a nitrogen filled glove box (without air exposure) [27]. Under these circumstances no resistive switching was recorded. This changed after short air exposure (5 min.) of the samples: the cells went to a native low conductive state (the OFF state) and could be repeatably switched from to a high conductive ON state and back [27]. This experiment clearly proved that oxidation of the top contact metal is required for bistable electrical switching of this kind of memory elements. Additional proof was recently published by Hefczyc et al. [23] reporting experiments for horizontal devices with Au and Al electrodes. Whereas the two memory cells (I and II) of a structure with common Au electrode were completely independent upon each other (Fig. 5a), both cells of the inverse structure with common Al electrode (Fig. 5b) were always in the same state [23]. This experiment clearly indicated that the bipolar resistive switching was localized at the CuTCNQ\Al interface and was thus not a bulk phenomenon.

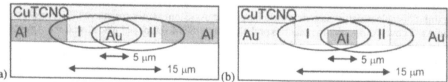

Figure 5. Scheme of horizontal Au\CuTCNQ\Al devices with (a) common Au electrode and (b) common Al electrode (redrawn and modified from [23]).

Further experiments with a heat sensitive infrared camera on Cu\CuTCNQ\Al cross-bar memory cells (prepared as reported in [28] and previously switched to the ON state) allowed us to visualize the presence of conductive channels within the memory elements [27]. Current-time experiments on Au\CuTCNQ nanowire\Al memory cells (in air) also showed a signature for filamentary growth. As depicted in Figure 6, the current increased (in absolute value) from the beginning of the experiment by sudden jumps which might be attributed to filament growth [29].

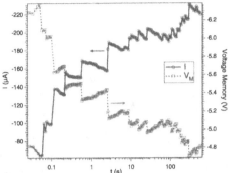

Figure 6. I-t and V_M-t curves for an Au\CuTCNQ nanowire\Al memory cell (200 μm by 200 μm) switched to the ON state by applying a constant voltage (-7 V) to the memory element in series with a 10 kΩ load resistor [29].

In order to develop a model for the resistive switching of CuTCNQ based memories we began to compare their behavior with the ones reported in the literature for other kinds of memory cells. The polarity dependence of CuTCNQ ON/OFF resistive switching led us to consider electrochemistry based switching mechanisms, as already been experimentally evidenced in 1980 for the CuTCNQ analog CuTNAP (were TNAP denotes 11,11,12,12-tetracyano-2,6-napthoquinodimethane) [18] and more recently suggested for CuTCNQ [9,26]. Within the literature two different categories of electrochemistry based memories were found which involved Cu cations: the programmable metallization cell (PMC) [5], and copper sulfide [6]. Both of them exhibit bipolar resistive electrical switching and in both cases the switching is believed to happen within the memory material itself: the glassy electrolyte containing Cu cations in the case of the PMC [5] and the Cu^+ cations within the copper sulfide [6]. A further analogy with metal-TCNQ charge-transfer complexes is that resistive switching was also reported for AgTCNQ [30], Ag cations in the PMC [31], as well as silver chalcogenides [32]. These last compounds behave as solid ionic conductors in which the high conductivity (1-4 $S.cm^{-1}$) is related to mobile Ag^+ cations [33]. Terabe et al. have shown that these compounds exhibit bistable resistive electrical switching in Ag\Ag$_2$S\nm gap\Pt structures [32], consisting of an Ag$_2$S covered Ag wire mounted as probe tip in a c-STM setup, and placed with the Ag$_2$S layer a few nm away of a Pt substrate. Negative polarization of the Pt contact with respect to the Ag contact led to growth of silver nanoclusters from the Ag$_2$S towards the Pt, bridging the gap and decreasing the device resistance. The growth of the nanoclusters was explained by electrochemical reduction of mobile Ag^+ cations originating from the Ag$_2$S solid ionic conductor. Applying the opposite polarity to the structure led to dissolution of the Ag nanoclusters by electrochemical oxidation, so that the device resistance increased again.

It has been reported that CuTCNQ (phase I) has a conductivity of 0.25 S.cm^{-1} [22] which is very similar to those of the silver chalcogenide solid ionic conductors [33]. Furthermore, Yamaguchi et al. [21] have shown that conductive hillocks of 40 nm height can be formed by applying negative polarization (-2V) to a Pt/Ir conductive scanning tunneling microscopy (c-STM) tip above a CuTCNQ layer. These hillocks, which were stable for over 2 days, could be switched back by inversing the polarity of the signal applied to the c-STM tip (+2 V). This experiment is very similar to the one of Terabe et al. reporting the growth of Ag nanoclusters at the end of an Ag\Ag$_2$S c-STM tip a few nanometers on top of a Pt substrate [32]. Therefore it can be assumed that the conductive hillocks observed by Yamaguchi et al. [21] on the CuTCNQ substrate are made of metallic copper.

This interpretation allows explaining the bipolar resistive electrical switching of CuTCNQ through nanometer sized gaps (using for example c-STM tips as top contact), but not the switching of CuTCNQ based memory elements with easily oxidizable metal top contacts (Al, Yb, Ti, ...). The answer to this open question was found by experiments published by Serebrennikova and White [34] on defect sites in native alumina films. In fact, the authors proved the presence of microscopic defect sites with radii between 1 and 10 μm in native aluminum oxide films (2-3 nm thickness) by scanning electrochemical microscopy (SECM). Since the resolution of the SECM was restricted by the diameter of the SECM tip (9 μm) the existence of much smaller defects is highly likeable. Similar experiments performed by the same group [35] also indicated the presence of defect sites ranging from 0.1 and 50 μm for Ti, Ta, and Al electrodes covered by their respective nominal oxide (TiO$_2$, Ta$_2$O$_5$, and Al$_2$O$_3$). Since we observed bipolar resistive electrical switching of CuTCNQ memories with (partially oxidized) Al and Ti top contacts we ascribed the switching to the presence of defect sites within the oxide layer. Because the defect sites are much more conductive than the oxide layer (Al$_2$O$_3$, TiO$_2$, ...) [35] it is reasonable to assume that they originate from tiny cracks within the oxide layer, or from (metallic) impurities which are not oxidized in air. For this reason we consider that the oxide layer just acts as a porous layer in which nanometer size gaps are formed in which the switching occurs by growth of Cu filaments. Preliminary experiments performed on Si\n+ Si\SiO$_2$ (5nm)\CuTCNQ\Au memory cells seem to confirm this hypothesis of switching inside an ion-permeable layer [36]. In fact, bipolar resistive electrical switching was also observed for this kind of structure (Fig. 7). Compared to the typical Cu\CuTCNQ\Al devices (with native Al$_2$O$_3$ between the CuTCNQ and the Al top contact), Si\n+ Si\SiO$_2$\CuTCNQ\Au memories switch with opposite polarity, since for these memories the oxide layer is situated between the CuTCNQ and the bottom contact.

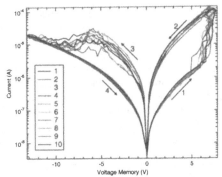

Figure 7. Ten consecutive log|I|-V curves of a Si\n+Si\SiO₂ (5 nm)\CuTCNQ\Au memory cell. The voltage sweep (+14V to -13V) was applied to the Au top contact (through a load resistor of 47 kΩ) [36].

Taking into account all these information we propose the following model for the bipolar resistive electrical switching of CuTCNQ based memories in presence of a porous (i.e. ion-permeable) layer (for example an oxide layer) or a nanometer sized gap (Fig. 8).

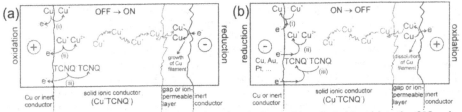

Figure 8. Model of CuTCNQ bipolar resistive electrical switching. Switching (a) to the ON and (b) to the OFF state. The ion-permeable layer can be formed by a native oxide (from for example an oxidizable Al, Yb, or Ti electrode), or deposited under controlled conditions.

In this mechanism, the switching device contains two relevant elements: the solid ionic conductor (CuTCNQ) with mobile Cu^+ ions, and an ion-permeable "switching layer", for example a porous (oxide) layer. By applying a sufficient negative voltage to the electrode contacting the porous layer, mobile Cu^+ ions from the solid ionic conductor are electrochemically reduced to metallic copper, forming a conductive channel (growing from the cathode towards the anode) which bridges the nanometer sized gap. In this way, the resistance across the gap is significantly decreased, so that the memory element goes from the OFF state to the ON state. During the reduction of Cu^+ cations at the negatively polarized electrode, an oxidation reaction takes place at the other electrode. This reaction can be (i) the oxidation of metallic Cu into Cu^+ in the case of a copper electrode, (ii) the oxidation of Cu^+ (originating from CuTCNQ) into Cu^{2+}, or (iii) the oxidation of TCNQ⁻ (from CuTCNQ) into neutral TCNQ. Whereas reaction (i) can only occur with metallic Cu electrodes, both reactions (ii) and (iii) are also possible for Cu and other electrodes. Since the standard electrode potentials of all involved electrochemical couples are not known for solid-state reactions it is not possible to foresee which of the 3 reactions prevails. Taking into account that equation (i) would require Cu atoms to leave the structure of the metal and that bipolar resistive switching has been observed for memory cells with non-Cu bottom contacts [17], reactions (ii) and (iii) are more likely to occur. This means that even if reaction (ii) would be favored towards reaction (iii), after some time, when the diffusion layer at the metal\CuTCNQ interface becomes depleted of Cu^+ cations, the oxidation of $TCNQ^-$ into neutral TCNQ (reaction iii) has to occur. Therefore our model also allows explaining the detection of neutral TCNQ in the RAMAN spectra of CuTCNQ memories programmed for an extended time to the ON state [19].

Switching the memory elements back to the OFF state is performed by applying the opposite polarity. When a sufficient positive voltage is applied to the metal electrode in contact with the "switching layer" the copper filament within the ion-permeable layer or nanometer sized gap is electrochemically oxidized back into Cu^+ cations, which migrate within the solid ionic conductor CuTCNQ. Since the "switching layer" is no more bridged by a conductive channel, the memory cell goes back to the OFF state. Simultaneously the reduction of the previously formed species Cu^+ (i), Cu^{2+} (ii), or TCNQ (iii) will occur at the other, negatively polarized electrode. It is also noteworthy to mention that the dissolution of the copper filament might not be complete, depending upon the experimental conditions. This can explain why the OFF state reading current often increases after the first switch to the ON state.

Kever et al. already previously suggested that the resistive switching of CuTCNQ based memories with Al top contacts was due to "Cu-ion based electrochemical switching" [26]. They further compared switching in a Cu\CuTCNQ\nativeAlOx\Al cell with that in an Al\UV-O₂ formed AlOx\Cu cell. The observation of "similar" bipolar switching for the latter cell (with same polarity dependence towards electrode materials) was taken as further proof of their model in which aluminum oxide represents the "electrolyte" and the CuTCNQ layer appears to be nothing but "a suitable spacer, which possibly stabilizes the reversible switching by acting as a Cu ion buffer", but would not constitute an a-priory required layer. While this experiment indeed elucidates the crucial role of the AlOx switching layer, a detailed comparison of the switching I-V characteristics shows important differences between the two kinds of structures, (for example a different switching stability) – indicating the presence of CuTCNQ *does* play an important role.

In our model the CuTCNQ clearly participates to the switching process. It acts as "solid ionic conductor" providing the mobile Cu^+ ions required for the electrochemical switching. This property might also explain why other Cu^+ salts (without mobile Cu^+ ions) as CuI at room temperature are not suitable for bistable resistive switching memories [37]. Furthermore the fact that resistive electrical switching was observed in recent experiments with SiO_2 layers (Fig. 7, [36]) is a clear indication that the relevant role of the "switching" layer is its permeability to the ions.

As described within the previous paragraphs, all experiments having previously served as evidence for bulk switching (Fig. 3) of CuTCNQ based memories (with oxide layer or a nanometer sized gap) can also be explained by an interfacial effect. Furthermore, our model can be generalized to other Cu^+ and Ag^+ cation based charge-transfer complexes behaving as solid state ionic conductors, as for example CuTNAP [18], AgTNAP [18], and CuDDQ (2,3-dichloro-5,6-dicyano-p-benzoquinone) [38]. As shown in Figure 9, the switching system implies two essential components: a solid ionic conductor $M_x^+A_y^{z-}$ (with *monovalent* mobile M^+ cations, for example Cu^+ and Ag^+, but probably also the highly toxic Tl^+) and a "switching layer" in which conductive channels of the M metal are formed by electrochemical reduction of the M^+ cations and dissolved by electrochemical oxidation. This switching layer should be permeable to the M^+ cations for example containing pores or nanometer sized gaps, through which the M^+ cations can move by migration and diffusion, and in which the conductive channels build up and dissolve. This general model also holds for the switching of *inorganic* solid ionic conductors with an ion-permeable layer or gap, as for example Ag_2S [32], or with an oxide switching layer as the Cu-Te/GdOx bilayer structure by Aratani et al. [39].

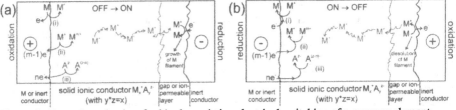

Figure 9. General mechanism for bipolar resistive electrical switching of a memory element composed of a solid ionic conductor and an ion-permeable layer or gap. Switching (a) to the ON and (b) to the OFF state. The ion-permeable layer can be formed by a native oxide (from for example an oxidizable Al, Yb, or Ti electrode), or deposited under controlled conditions (then the contact can also be an inert conductor).

As previously mentioned, our model was developed for memory systems composed of a solid ionic conductor and an ion-permeable "switching layer". Some papers also reported resistive electrical switching of solid ionic conductors without this "switching layer", for example by Cu filament growth in Cu_2S [6]. Similar switching might also be possible for Cu^+ and Ag^+ metal-organic charge transfer complex based solid ionic conductors as for example CuTCNQ in absence of an ion-permeable "switching layer", or for a "switching layer" already bridged by several conductive channels. Indeed, this may be related to the two different "ON" states observed in 50 to 300 nm thick CuTCNQ cells [40]. Since the filament growth within the solid ionic conductor likely goes ahead with (major) structural changes inside the material it can be expected that corresponding memory devices might – depending upon the material - suffer from low endurance and retention time (the 2nd, more conductive "metallic" ON state in [40] was indeed reported to be unstable).

CONCLUSIONS

In this work we gave strong evidence that the bipolar resistive electrical switching of CuTCNQ based memories with aluminum top contacts or with Pt/Ir c-STM tips as top contacts occurred by an interfacial effect and not by a bulk effect. We developed a model of the switching process implying two different layers: a solid-state ionic conductor (CuTCNQ), acting as source of mobile Cu^+ ions, and a "switching layer". This "switching layer" was depicted in our model only as simple ion-permeable spacer between the CuTCNQ layer and a conductive electrode, with nanometer sized pores or gaps in which conductive Cu channels can be formed by electrochemical reduction of Cu^+ cations, and afterwards dissolved by electrochemical oxidation. Previously published "evidence" for bulk switching of CuTCNQ was also interpreted by this mechanism in terms of interfacial switching. Generalization of the switching mechanism to other Cu^+ and Ag^+ based charge-transfer complexes behaving as solid state ionic conductors (including inorganic materials) was also presented.

Further studies aiming improvement of the bipolar resistive electrical switching properties of Cu^+ and Ag^+ based charge-transfer complex will require optimizing the properties of the solid ionic conductor, of the "switching layer", and the interface between these two materials.

Since localized switching of CuTCNQ with c-STM tips was successful, and integration of the material was possible in 100 nm diameter vias of Cu CMOS BEOL wafers, future integration experiments with an optimized "switching layer" and an improved interface between the "switching layer" and the solid ionic conductor, can be expected to be very promising for nonvolatile memory applications.

ACKNOWLEDGMENTS

This research was performed within the framework of the NOSCE MEMORIAS (FP6-507934) and EMMA (FP6-IST-Strep 3375) projects of the European Commission.

REFERENCES

At the date this paper was written, URLs or links referenced herein were deemed to be useful supplementary material to this paper. Neither the author nor the Materials Research Society warrants or assumes liability for the content or availability of URLs referenced in this paper.

1. http://www.itrs.net/Links/2006Update/2006UpdateFinal.htm (last accessed on 7th April 2008)
2. I.G. Baek, M.S. Lee, S. Seo, M.J. Lee, D.H. Seo, D.S. Suh, J.C. Park, S.O. Park, H.S. Kim, I.K. Yoo, U.I. Ching, and J.T. Moon, *Electron Devices Meeting, 2004. IEDM Technical Digest. IEEE International*, 587 (2004).
3. S. Hudgens and B. Johnson, *MRS Bull.* 29, 829 (2004)
4. M.N. Kozicki, M. Park, and M. Mitkova, *IEEE Trans. Nanotechnol.* 4, 331 (2005).
5. M.N. Kozicki, M. Balakrishnan, C. Gopalan, C. Ratnakumar, and M. Mitkova, *Non-Volatile Mem. Technol. Symp.* (2005), 83.
6. S. Kaeriyama, T. Sakamoto, H. Sunamura, M. Mizuno, H. Kawaura, T. Hasegawa, K. Terabe, T. Nakayama, and M. Aono, *IEEE J Solid-State Circ.* 40, 168 (2005).
7. G. Meijer, *Science* 319, 1625 (2008).
8. Y. Yang, L. Ma, and J. Wu, *MRS Bull.* 29, 833 (2004).
9. J.C. Scott and L.D. Bozano, *Adv. Mat.* 19, 1452 (2007).
10. Q.-D. Ling, D.-J. Liaw, E. Y.-H. Teo, C. Zhu, D. S.-H. Chan, E.-T. Kang, and K.-G. Neoh, *Polymer* 48, 5182 (2007).
11. R.S. Potember, T.O. Poehler, and D.O. Cowan, *Appl. Phys. Lett.* 34, 405 (1979).

12. T. Kever, C. Nauenheim, U. Böttger, and R. Waser, *Thin Solid Films* 515, 1893 (2006).
13. R. Müller, J. Billen, R. Naulaerts, O. Rouault, L. Goux, D.J. Wouters, J. Genoe, and P. Heremans, *Mater. Res. Soc. Symp. Proc.* 997, I01-10 (2007)
14. R. Müller, S. De Jonge, K. Myny, D.J. Wouters, J. Genoe, and P. Heremans, *Appl. Phys. Lett.* 89, 223501 (2006).
15. R. Müller, L. Goux, D.J. Wouters, J. Genoe, and P. Heremans (unpublished)
16. T. Oyamada, H. Tanaka, K. Matsushige, H. Sasabe, and C. Adachi, *Appl. Phys. Lett.* 83, 1252 (2003).
17. R. Müller, R. Naulaerts, J. Billen, J. Genoe, and P. Heremans, *Appl. Phys. Lett.* 90, 063503 (2007).
18. R.S. Potember, T.O. Poehler, A. Rappa, D.O. Cowan, and A.N. Bloch, *J. Am. Chem. Soc.* 102, 3659 (1980)
19. E.I. Kamitsos, C.H. Tzinis, and W.M. Risen, Jr., *Solid State Commun.* 42, 561 (1982).
20. M. Matsumoto, Y. Nishio, H. Tachibana, T. Nakamura, Y. Kawabata, H. Samura, and T. Nagamura, *Chem. Lett.* 1021 (1991).
21. S. Yamaguchi and R.S. Potember, *Mol. Cryst. Liq. Cryst.* 267, 241 (1995).
22. R.A. Heintz, H. Zhao, X. Ouyang, G. Grandinetti, J. Cowen, K.R. Dunbar, *Inorg. Chem.* 38, 144 (1999).
23. A. Hefczyc, L. Beckmann, E. Becker, H.-H. Johannes, and W. Kowalsky, *Phys. Stat. Sol. (a)* 205, 647 (2008).
24. C. Sato, S. Wakamatsu, K. Tadokoro, and K. Ishii, *J. Appl. Phys.* 68, 6535 (1990).
25. J.J. Hoagland, X.D. Wang, and K.W. Hipps, *Chem. Mater.* 5, 54 (1993).
26. T. Kever, U. Böttger, C. Schindler, and R. Waser, *Appl. Phys. Lett.* 91, 083506 (2007).
27. J. Billen, S. Steudel, R. Müller, J. Genoe, and P. Heremans, *Appl. Phys. Lett.* 91, 263507 (2007).
28 J. Billen, R. Müller, J. Genoe, and P. Heremans, *Proceed. 2nd Int. Conf. Mem. Technol. & Des.*, Giens (France) (2007) 135.
29. R. Müller, R. Naulaerts, J. Genoe, and P. Heremans, (unpublished).
30. Z.Y. Fan, X.L. Mo, G.R. Chen, and J.G. Lu, Rev. *Adv. Mater. Sci.* 5, 72 (2003).
31. M.N. Kozicki, M. Mitkova, M. Park, M. Balakrishnan, and C. Gopalan, *Superlat. & Microstruct.* 34, 459 (2003).
32. K. Terabe, T. Hasegawa, T. Nakayama, and M. Aono, *RIKEN Rev.* 37, 7 (2001).
33. T. Ohachi and I. Taniguchi, *J. Cryst. Growth* 13-14, 191 (1972).
34. I. Serebrennikova and H.S. White, *Electrochem. Sol.-State Lett.* 4, B4 (2001).
35. I. Serebrennikova, S. Lee, and H.S. White, *Far. Discuss.* 121, 199 (2002).
36. R. Müller, C. Krebs, L. Goux, D.J. Wouters, J. Genoe, and P. Heremans (unpublished).
37. R. Müller, J. Genoe, and P. Heremans, (unpublished)
38. R.T. Weitz, A. Walter, R. Engl, R. Sezi, and C. Dehm, *Nano Lett.* 6, 2810 (2006).
39. K. Aratani, K. Ohba, T. Mizuguchi, S. Yasuda, T. Shiimoto, T. Tsushima, T. Sone, K. Endo, A. Kouchiyama, S. Sasaki, A. Maesaka, N. Yamada, and H. Narisawa, *Electron Device Meeting (IEDM)*, 783 (2007)
40. T. Kever, B. Klopstra, U. Böttger, and R. Waser, 7th *Ann. Non-Volatile Mem. Technol. Symp.*, 119 (2006)

Mater. Res. Soc. Symp. Proc. Vol. 1071 © 2008 Materials Research Society 1071-F05-01

Molecular Conformation-Dependent Memory Effects in Non-Conjugated Polymers With Pendant Carbazole Moieties

Siew Lay Lim[1,2], Qidan Ling[2], Eric Yeow Hwee Teo[3], Chun Xiang Zhu[3], Daniel Siu Hung Chan[3], En Tang Kang[2], and Koon Gee Neoh[2]
[1]NUS Graduate School of Integrative Sciences and Engineering (NGS), National University of Singapore, Singapore
[2]Department of Chemical and Biomolecular Engineering, National University of Singapore, Singapore
[3]SNDL, Department of Electrical and Computer Engineering, National University of Singapore, Singapore

ABSTRACT

Single-layer devices of the structure indium-tin-oxide/polymer/aluminum were fabricated from two non-conjugated polymers with pendant carbazole groups in different spacer units. The device based on poly(2-(N-carbazolyl)ethyl methacrylate) (PMCz) exhibited non-volatile write-once-read-many-times (WORM) memory behavior with an ON/OFF current ratio up to 10^6, while the device based on poly(9-(2-((4-vinylbenzyl)oxy)ethyl)-9H-carbazole) (PVBCz) exhibited volatile memory behavior with an ON/OFF current ratio of approximately 10^3. In the absence of a spacer unit between the pendant carbazole group and the main chain, regioregular poly(N-vinylcarbazole) (PVK) exhibited only one conductivity state (ON state). The formation of carbazole excimers resulting from conformation-induced conductance switching under an electric field was revealed $in situ$ by fluorescence spectroscopy. The electrical behavior of the polymer in a device was found to be dictated by the chemical structure and steric effect of the spacer unit between the pendant carbazole group and the main chain.

INTRODUCTION

Organic materials are promising candidates for future molecular-scale device applications in new information storage technologies as the need to overcome the potentially limiting scaling difficulties present in the semiconductor industries intensifies [1]. In particular, polymer memories have emerged as an active research topic in recent years. Polymer materials possess unique properties, such as good mechanical strength and flexibility, as well as the possibility for molecular design through chemical synthesis. Polymer memory devices are easily fabricated by solution-processing methods, such as spin-coating and ink-jet printing, and a variety of materials, including plastics and metal foils, can be used as substrates for deposition of the polymers. The memory cells can be addressed by two-terminal structures and stacked to form high-density, three-dimensional (3-D) data storage devices. Polymer memories can potentially provide low-cost, low-power, flexible, and light-weight devices with high-capacity data storage [2].

Rather than encoding "0" and "1" from the number of charges stored in a cell in silicon devices, a polymer memory stores data, for example, on the basis of the high and low conductivity responses to an applied voltage [3]. Although conductance switching due to charge-transfer, electro-reduction, charge-tunneling and others has been widely reported [4-6], conformation-induced conductance switching for polymer memory device application has yet to be explored in detail. Here, we illustrate the effect of chemical structure on the conformation-induced switching properties of two non-conjugated polymers with pendant carbazole groups in different spacer units. Poly(N-vinylcarbazole) (PVK) is used as a reference. The voltage-induced conformation changes in the polymer films were revealed via *in-situ* fluorescence spectroscopy.

EXPERIMENT

All polymers used were synthesized by free-radical homopolymerization of the respective monomers in dry tetrahydrofuran (THF) for 72 h at 65°C, with 1 mol/L monomer feed concentration and azobisisobutyronitrile as the initiator, under an argon atmosphere. Each polymer was precipitated in methanol and purified by Soxhlet extraction with methanol. Poly(N-vinylcarbazole) (PVK) was synthesized by polymerization of 9-vinylcarbazole (98%). [GPC: M_w ~ 12,000 g mol^{-1}, PDI ~ 1.5]. 2-(N-carbazolyl)ethyl methacrylate and the corresponding polymer, PMCz, were synthesized as reported [7]. [GPC: M_w ~ 15,800 g mol^{-1}, PDI ~ 2.7] For the preparation of 9-(2-((4-vinylbenzyl)oxy)ethyl)-9H-carbazole monomer,[31] 9H-carbazole-9-ethanol (97%, 5.6 g or 25.2 mmol) and sodium hydride (95%, 1.0 g or 41.7 mmol) were dissolved in THF. 4-Vinylbenzyl chloride (97%, 4.0 ml or 25.6 mmol) was added in one portion and the solution was stirred in a 60°C oil bath for 24 h under an argon atmosphere. The excess solvent was evaporated and the product extracted by diethyl ether. The corresponding polymer, PVBCz, was then prepared by free-radical polymerization as described above. [GPC: M_w ~ 24,000 g mol^{-1}, PDI ~ 1.7].

Indium-tin oxide (ITO)-coated glass substrates of 1 cm × 2 cm in size were pre-cleaned by ultrasonication with water, acetone and isopropanol, in that order. The polymer memory devices were fabricated by spin-coating the polymer films of about 50 nm in thickness on ITO from 10 mg/ml solutions of PVK in toluene, and of PMCz and PVBCz in N,N'-dimethylacetamide, followed by solvent removal in a vacuum chamber at 10^{-5} Torr and 60°C for 12 h. Thermal evaporation of the aluminum electrodes of about 0.16, 0.04 and 0.0225 mm^2 in areas and 0.1 μm in thickness, through a shadow mask, on the polymer film produced the ITO/polymer/Al devices. All electrical characterizations were carried out using a HP 4156A semiconductor parameter analyzer equipped with an Agilent 16440A SMU/pulse generator under ambient conditions. ITO was maintained as the ground electrode. Fluorescence spectroscopy was performed *in situ* on a Shimadzu RF-5301PC Spectrofluorophotometer, with the external bias across the two electrodes imposed by a HP 6282A DC voltage source.

DISCUSSION

The chemical structures of the three polymers with pendant carbazole groups and the schematic diagram of the indium-tin oxide/polymer/aluminium (ITO/polymer/Al) device are shown in Figure 1(a) and Figure 1(b), respectively.

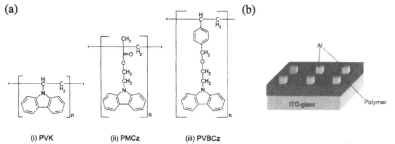

(i) PVK   (ii) PMCz   (iii) PVBCz

Figure 1. (a) Molecular structures of poly(N-vinylcarbazole) (PVK), poly(2-(N-carbazolyl)ethyl methacrylate (PMCz) and poly(9-(2-((4-vinylbenzyl)oxy)ethyl)-9H-carbazole (PVBCz). (b) Schemetic diagram of the memory device consisting of a thin film (~50 nm) of the polymer sandwiched between an indium-tin-oxide (ITO) substrate and an aluminum top electrode.

The current density-voltage (J-V) curve of the device fabricated with PVK as the active polymer layer shows the device to be always in a single high conductivity state in the working voltage range (see Figure 2(a)). In comparison, devices fabricated with PMCz and PVBCz (Figures 2(b) and 2(c), respectively) exhibit conductance switching associated with the electrical bistability [7,8]. The ITO/PMCz/Al device switches from the low-conductivity (OFF) state to the high-conductivity (ON) state at about -1.8 V when the voltage applied was increased from 0 V to -2 V (Sweep 1 in Figure 2(b)). The ON-state is retained in the subsequent voltage sweeps in the negative (Sweep 2 in Figure 2(b)) or positive bias (not shown). The device remains in the ON-state even after the power has been turned off, with an ON/OFF current ratio up to 10^6 when read at -1 V. The ITO/PMCz/Al device thus exhibits write-once-read-many-times (WORM) memory behavior, with both OFF and ON states stable under a constant voltage stress of -1.0 V and for up to 10^8 read cycles at -1.0 V (pulse width = 1 μs, pulse period = 2μs).

The ITO/PVBCz/Al device switches from the OFF to ON state at about -2.0 V, with an ON/OFF current ratio of about 10^3, when a voltage sweep from 0 V to -3 V was applied (Sweep 1 in Figure 2(c)). The device remains in the ON-state when the voltage sweep was repeated (Sweep 2). As shown by Sweep 3 in Figure 2(c), the "written" ON-state of the memory device cannot be erased by applying reverse bias, indicating that the device is non-rewritable. However, the ON state is retained only for a period of 2 to 10 min after turning off the power, after which the device can be "written" again by applying the appropriate voltage (Sweep 4). The ON state is still volatile even when a larger (but non-degrading) bias is applied during Sweep 1. Although the memory effect of the device is volatile, the ON-state can be electrically sustained by refreshing pulses of -1 V (pulse width = 10 ms) every 5 s (Figure 2(c)), or under a continuous bias of -1 V, while maintaining a ON/OFF current ratio of approximately 10^3. Therefore, the behavior of the ITO/PVBCz/Al device shares some similarities with that of a static random access memory (SRAM), except for the relatively long retention time (≥ 2 min) of the ON state. No resistance degradation is observed for both the ON and OFF states during read cycles testing up to 10^8 read cycles.

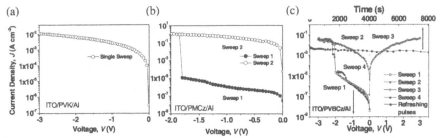

Figure 2. *J-V* characteristics of (a) an ITO/PVK/Al device showing a single conductivity state, (b) and (c) an ITO/PMCz/Al device and ITO/PVBCz/Al device, respectively, in the OFF- and ON-state, with the corresponding OFF-to-ON transitions at -1.8 V and -2.0 V ((c) also shows the ON-state being maintained by refreshing pulses).

The carbazole group is an electron-donor and hole-transporter and has a tendency to form a partial or full face-to-face conformation with the neighboring carbazole groups to result in extended electron delocalization [9]. Under negative bias, hole injection from ITO oxidizes the carbazole groups near the interface, forming positively charged species. As an effective donor, nearby neutral carbazole groups undergo charge transfer interactions with the positively charged carbazole groups, thus inducing the conformation change and delocalizing the positive charge to neighboring groups. Such regions of electron delocalization provide pathways for charge carrier hopping via the carbazole groups in the direction of the electric field. When the power is turned off, some residual positive charges are localized on the carbazole groups. As shown by molecular mechanics simulation, the carbazole pendant groups attached to the main chain via flexible C-O linkages in PMCz and PVBCz are initially in random orientations (Figure 3). The switching effect in the PMCz and PVBCz devices probably arises from a change in conformation of the polymers via rotations of the carbazole groups to result in a more regioregular arrangement [8], similar to that of PVK (see Figure 3). The introduction of flexible spacers between the carbazole pendant groups and the main chain of the polymer results in a lowering of the glass transitions temperatures (T_g) in PMCz and PVBCz and increased rotational flexibility [10].

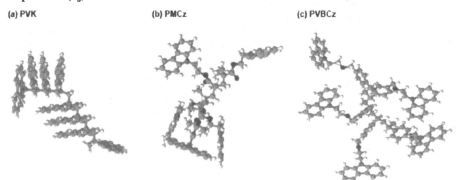

Figure 3. Simulated 3D models by molecular mechanics showing the optimized geometry corresponding to the minimum energy states in (a) PVK, (b) PMCz and (c) PVBCz.

In situ fluorescence spectroscopic measurements of the polymer in the devices (through the transparent ITO back contact) showed increases in emission intensities at 380 nm and 420 nm, due to the formation of the partially eclipsed carbazole excimers and the totally-eclipsed sandwich-like excimers [11], respectively, under the influence of electric field (Figure 4). The broadening of the emission peak at about 360 nm, attributed to monomer fluorescence of the carbazole chromophore, also indicates an increase in the types of emitting species in addition to the monomeric chromophores. Consistent with the volatile nature of the PVBCz memory device, the fluorescence emission spectrum obtained 10 min after power-off shows only the monomeric emission [8]. In contrast, the emission spectrum of the PMCz device remains unchanged (in the excimer state) when measured 10 min after turning off the power, in accordance with the non-volatile nature of the memory device.

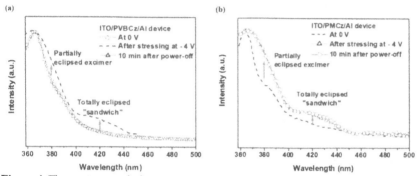

Figure 4. Fluorescence emission spectra showing changes in intensity at ~380 nm and ~420 nm of the (a) ITO/PVBCz/Al and (b) ITO/PMCz/Al devices at 0 V, after applying a voltage bias, and after power-off.

Comparing the chemical structures and memory properties of PMCz and PVBCz, the bulkier spacer between the pendant carbazole group and the backbone in PVBCz allows a larger free volume and, hence, a greater degree of conformational freedom for relaxation through, for example, rotation of the carbazole and phenyl rings about the C-O bond, causing the ON-state to be unstable without the continuous application of a bias [8]. Without a continually applied external voltage, the carbazole groups in PVBCz return to their original random conformation (OFF-state). On the other hand, the neighboring electron-withdrawing O-C=O groups in PMCz further stabilize the positively charged carbazole groups, prolonging the retention of the ON-state. The more amorphous nature of PVBCz, compared to PMCz, also means that the carbazole groups are further apart in the ground state. The charge transport pathways formed by conformational changes may therefore be shorter or less continuous, resulting in a lower ON state current in the PVBCz device. Thus, in contrast to the WORM-type behavior exhibited by the PMCz device, the memory effect exhibited by the PVBCz device is volatile.

CONCLUSIONS

Non-volatile and volatile conductance switching effects are observed in single-layer memory devices fabricated from non-conjugated polymers with conformationally disordered pendant carbazole groups in ethylacrylate and benzyloxyethyl spacer units, respectively. Both polymer memory devices operate via the electric-field-induced conformation ordering of the pendant carbazole groups in the polymer films. The formation of carbazole excimers due to such conformation changes are observed via *in-situ* fluorescence spectroscopy. The extents of regio-regularity, conformation ordering, and conformational relaxation, in turn, are dictated by the chemical structure and steric effect of the spacer units between the carbazole moiety and the main chain.

ACKNOWLEDGMENTS

One of authors (S. L. Lim) is supported by A*STAR Graduate Scholarship (AGS), administered by Agency for Science, Technology and Research (A*STAR), Singapore.

REFERENCES

1. S. R. Forrest, *Nature* **428**, 911 (2004).
2. Q. D. Ling, D. J. Liaw, E. Y. H. Teo, C. X. Zhu, D. S. H. Chan, E. T. Kang and K. G. Neoh, *Polymer* **48**, 5182 (2007).
3. A. Stikeman, *Technol. Rev.* **105**, 31 (2002).
4. A. Bandyopadhyay, and A. J. Pal, *Chem. Phys. Lett.* **371**, 86 (2003).
5. J. Ouyang, C. W. Chu, C. R. Szmanda, L. Ma, and Y. Yang, *Nat. Mater.* **3**, 918 (2004).
6. S. Patil, Q. X. Lai, F. Marchioni, M. Y. Jung, Z. H. Zhu, Y. Chen, F. Wudl, *J. Mater. Chem.* **16**, 4160 (2006).
7. E. Y. H. Teo, Q. D. Ling, Y. Song, Y. P. Tan, W. Wang, E. T. Kang, D. S. H. Chan, C.X. Zhu, *Org. Electron.* **7**, 173 (2006).
8. S. L. Lim, Q. D. Ling, E. Y. H. Teo, C. X. Zhu, D. S. H. Chan, E. T. Kang and K. G. Neoh, *Chem. Mater.* **19**, 5148 (2007).
9. J. Vandendriessche, P. Palmans, S. Topet, N. Boens, F. C. De Schryver, H. Masuhara, *J. Am. Chem. Soc.* **106**, 8057 (1994).
10. Y. W. Chen, B. Zhang, F. Wang, *Opt. Commun.* **228**, 341 (2003).
11. K. Davidson, I. Soutar, L. Swanson, J. Yin, *J. Polym. Sci. Pol. Phys.* **35**, 963 (1997).

Mater. Res. Soc. Symp. Proc. Vol. 1071 © 2008 Materials Research Society   1071-F05-11

Dependence on Organic Thickness of Electrical Characteristics Behavior in Low Molecular Organic Novolatile Memory

Yool Guk Kim[1], Sung Ho Seo[1], Gun Sub Lee[1], Jea Gun Park[1], and Jin Kyu Kim[2]

[1]Department of Electrical & Computer Engineering, Hanyang University, Nano SOI Process Laboratory, Room #101, HIT, 17 Haengdang-dong, Seoungdong-gu, Seoul, 133-791, Korea, Republic of

[2]Electron Microscopy Team, Korea Basic Science Institute, 113 Gwahangno (52 Eoeun-dong), Yusung-gu, Deajeon, 305-333, Korea, Republic of

ABSTRACT

We developed the devices to investigate the dependence on the organic thickness of electrical characteristics in small molecular organic nonvolatile memory. We developed four different thicknesses of organic layers, i.e., 30, 40, 50, and 100 nm, with a fixed middle layer thickness, were deposited using a high vacuum thermal evaporation. We confirmed that, as the organic layer thickness increases, the current level linearly decreases by an order of magnitude in a log-scale except for the 100-nm sample. The reason for this is that electron transfer occurs less frequently because of the decrease in the hopping frequency. Meanwhile, the switching characteristics did not much change. Therefore, we can conclude that the thickness of the organic layer does not significantly affect the switching characteristics except current level. In addition, it was confirmed that a 30-nm-thick organic layer was the best process condition for fabricating low-molecular organic nonvolatile memory.

INTRODUCTION

It has recently been reported that small molecular organic devices fabricated with a sandwich structure consisting of a top metal layer, small-molecular organic layer, middle metal layer, conductive organic layer, and a bottom metal layer demonstrated nonvolatile memory behaviors such as I_{on} (reading after programming)/I_{off} (reading after erasing) of greater than 1×10^1 and response times of ~10 ns. The organic conductive layers of these devices are 2-amino-4, 5-imidazoledicarbonitrile (AIDCN), Aluminum tris (8-hydroxyquinoline) (Alq3), and N,N'-bis(1-naphthyl)-1, 1'biphenyl4-4"diamine (α-NPD).[1~6] These small molecular organic nonvolatile memories require Al nanocrystals. However, we fabricated small molecular nonvolatile memories with Ni nanocrystals. The reason we chose Ni is that has a smaller grain boundary and a larger work function (~5.15 eV) that can make a deep quantum well in the energy band diagram. As a result, the memories showed a very large memory margin (I_{on} (reading after programming)/I_{off} (reading after erasing) ratio) and were more reproducible.

EXPERIMENTAL DETAILS

We developed a small molecular organic nonvolatile memory with the device structure of a bottom Al electrode, conductive organic layer (Alq3), Ni nanocrystals surrounded by NiO, upper conductive organic layer (Alq3), and a top Al electrode. Four different organic thicknesses, i.e., 30, 40, 50, and 100 nm, with a fixed middle layer thickness were deposited by using a high vacuum thermal evaporation, as shown in Figure. 1.

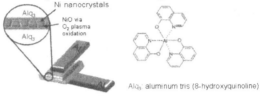

Figure 1. Device structure and a molecular formula of Alq3

To avoid contamination (e.g. H, O, N, Cl, and F atoms), all fabrication processes were carried out in the same evaporator without breaking the vacuum. The 80-nm-thick bottom Al electrode was thermally evaporated at a rate of 5 Å/s in a chamber pressure of 10^{-5} Pa using a first shadow mask with a line width of 3 mm. The 30-nm thick conductive organic layer (Alq3) was thermally evaporated at a rate of 1.0 Å/s using a 7 mm x 7 mm second mask. Next, the middle Ni nanocrystals layer was thermally evaporated at a rate of 0.1 Å/s using a 5 mm x 5 mm third shadow mask, and the wafer was transferred to the O_2 plasma chamber. To isolate the Ni nanocrystals from each other, we used an O_2 Plasma process to surround the Ni nanocrystals with NiO. Next, the second 30-nm-thick conductive organic layer (Alq3) was thermally evaporated at a rate of 1.0 Å/s using a 7 mm x 7 mm second mask. Finally, the 80 nm-thick top Al electrode was thermally evaporated at a rate of a 5 Å/s using a fourth mask, which was cross patterned against the bottom Al electrode. After fabrication, the sample of 3 mm x 3mm size evaluated the current versus voltage (I-V) characteristics. The sample for cross-sectional transmitted electron-microscopy (XTEM) observation was prepared using focused-ion-beam (FIB) etching. The crystal structure of the Ni nanocrystals was observed using ultra-high-voltage TEM (1.2 MV). The actual thickness was thinner than we originally assumed thickness, as shown in Figure 2.

Figure 2. XTEM images of small molecular organic nonvolatile memory cell:
(a) 30-nm sample, (b) 40-nm sample, (c) 50-nm sample, and (d) 100-nm sample.

Discussion

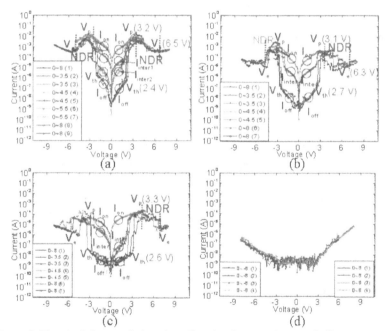

Figure 3. Electrical characteristics of small molecular organic nonvolatile memory cell:
(a) 30-nm I-V curves, (b) 40-nm I-V curves, (c) 50-nm I-V curves, and (d) 100-nm I-V curves

We first examined whether the devices described above had the potential for memory behavior. Figure 3(a) shows the I-V characteristics of our 30-nm small molecular nonvolatile memory cells. After the first applied voltage was swept from 0 to 10 V (called erase), as shown in line (1) in Figure 3(a), the second applied bias was swept from 0 to V_p or 3.5 V (called program), and then the current followed the high-resistance state (I_{off}). After the second applied bias (3.5 V: program voltage) was swept, as shown by line (2) in Figure 3(a), the third applied bias (reading after program) was again swept from 0 to V_p (3.5 V), as shown by line (3) in Figure 3(a), and then the current followed I_{on}. This shows that the device reads I_{off} (the low current state) at 1 V after erase (biasing 10 V above V_e). Otherwise, the device reads I_{on} (the high current state) at 1 V after program (biasing 3.5 V or V_p). Surprisingly, the ratio of I_{on} to I_{off} (memory margin) is $\approx 5 \times 10^3$, which is a sufficient current difference (bistable resistance difference) for multi level nonvolatile memory behavior. After the third applied bias (3.5 V: reading after program) was swept, as shown by line (3) in Figure 3(a), the fourth applied bias (4.5 V: first intermediate voltage) was swept from 0 to 4.5 V, as shown by line (4) in Figure 3(a), and then the current followed I_{on}. After the fourth applied bias, the fifth applied bias (4.5 V: reading after first intermediate) was swept from 0 to 5.5 V, as shown by line (5) in Figure 3(a), and then the current followed I_{inter1}. After the fifth applied bias, the sixth applied bias (5.5 V: second intermediate voltage) was swept form 0 to 5.5 V, shown by line (6) in Figure 3(a), and then current followed I_{inter1}. After the sixth applied bias, the seven applied bias (5.5 V: reading after second intermediate voltage) was swept from 0 to 5.5 V, as shown by line (7) in Figure 3(a), and then the current followed I_{inter2}. After the seventh applied bias, the eighth applied bias (8 V: erase voltage) was swept form 0 to 8 V, as shown by line (8) in Figure 3(a). Finally the ninth applied bias (8 V: reading after erase voltage) was swept form 0 to 8 V, as shown by line (9) in Figure 3(a), and then the current followed I_{off}. Like Figure 3, the 30-nm sample showed good electric behavior, but the 40-, 50-, and 100-nm samples did not. The Intermediate state was just one or none and I-V curve did not have a smooth shape as shown Figure 3(b), (c), and (d). Meanwhile, the switching characteristics of our device were V_{th} of ~2.5 V, V_p (program) of ~3.2 V, V_e (erase) of ~6.5 V, and I_{on} (after programming)/I_{off} (after erasing) of ~5×10^3 as shown Table 1.

Table 1. Electrical characteristics vs. small molecular thickness.

Thickness (nm)	V_{th} (V)	V_p (V)	V_e (V)
30	2.4	3.2	6.5
40	2.7	3.1	6.3
50	2.6	3.3	6.4
100	.	.	.

118

In addition, the interesting behavior of those characteristics was that they did not change much with varying organic thickness. However, the current level linearly decreases by an order of magnitude in a log-scale as the thickness increased by 10 nm. The reason is that electron transfer occurs less frequently because of the decrease in the hopping frequency. We fit the experimental results on the space charge-limited current (SCLC) mechanism in I_{off} (0 V~ V_{th}), as shown in Figure 4. As the space charges enhanced the internal electric field, the SCLC governed the current density (J) [7].

$$J = \frac{9\varepsilon\varepsilon_0 u V^2}{8L^3},$$

Where ε is the relative dielectric constant, ε_0 is the dielectric constant, and L is the sample organic layer thickness. Thus, the current increases as the square of the voltage increases in the SCLC mechanism. The experimental data governed by the SCLC mechanism if the slope became close to linear number 2 in log V vs. log I, as shown by Figure 4. It was shown us that the sample thickness became thicker when the slope became close to 2 except for the 100-nm sample. But we could not find a suitable mechanism in the other regions.

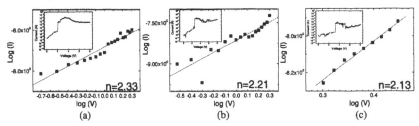

Figure 4. The scattered points are the experimental results, the solid line is the data fit on SCLC: (a) 30-nm sample, (b) 40-nm sample, and (c) 50-nm sample.

Conclusion

This study shows the dependence of the thickness of the organic layer on electrical characteristic behavior in small molecular organic nonvolatile memory. The interesting behavior of those characteristics is that they did not change much with varying thickness of the organic layer. We can conclude that the thickness of the organic layer does not significantly affect the switching characteristics except current level. Meanwhile, as the thickness of the organic layer increased, the device showed poor reproducibility and stability. Therefore, it was confirmed that a 30-nm-thick organic layer was the best process condition for fabricating small molecular organic nonvolatile memory.

ACKNOWLEDGMENTS

This research was supported by the Korea Ministry of Commerce, Industry and Energy for the 0.1 Terabit Non-volatile Memory Development.

Reference

1. L. P. Ma, J. Liu, and Y. Yang, *Appl. Phys. Lett.* **80**, 2997 (2002).

2. L. Ma, S. Pyo, J. Ouyang, Q. Xu, and Y. Yang, *Appl. Phys. Lett.* **82**, 1419 (2003).

3. L. D. Bozano, B. W. Kean, M. Beinhoff, K. R. Carter, P. M. Rice, and J. C. Scott, *Adv. Funct. Mater.* **15**, 1933 (2005).

4. J. He, L. Ma, J. Wu, and Y. Yang, *J. Appl. Phys.* **97**, 064507 (2005).

5. S. Pyo, L. Ma, J. He, Q. Xu, and Y. Yang, *J. Appl. Phys.* **98**, 54303 (2005).

6. L. D. Bozano, B. W. Kean, V. R. Deline, J. R. Salem, and J. C. Scott, *Appl. Phys. Lett.* **84**, 607 (2004).

7. S. M. Sze, "Physics of Semiconductor Devices," *MIS Diode and CCD*, 2Ed (Wiley, 1981) pp. 402-404.

Mater. Res. Soc. Symp. Proc. Vol. 1071 © 2008 Materials Research Society 1071-F05-13

Current Conduction Mechanism for Low-molecular Organic Nonvolatile Memory

Sung-ho Seo, Woo-sik Nam, Gon-sub Lee, and Jea-gun Park
Department of Computer Science and Engineering, Hanyang University, Nano SOI Process
Laboratory, Room # 101, HIT, Hanyang University, 17 Haengdang-dong, Seoungdong-gu,
Seoul, 133-791, Korea, Republic of

ABSTRACT

Organic devices fabricated with a top metal layer/conductive organic layer/middle metal layer/conductive organic layer/bottom metal layer structure have been reported to demonstrate nonvolatile memory behavior such as an after writing (I_{on})/after erasing (I_{off}) performance of > 1 x 10^1 and a response time of ~10 ns, when the organic conductive layers were AIDCN (*2-amino-4, 5-imidazoledicarbonitrile*), Alq3 (*Aluminum tris(8-hydroxyquinoline)*), or α-NPD. We fabricated an organic nonvolatile memory device with a structure of α-NPD/Al nanocrystals surrounded by Al_2O_3/α-NPD/Al, where α-NPD was *N,N'-bis(1-naphthyl)-1,1'biphenyl4-4''diamine*. A layer of Al nanocrystals, confirmed by a 1.25-MV high voltage transmission-electron-microscope, was uniformly produced between the α-NPD layers by Al layer evaporation at 1.0 Å/sec on the α-NPD followed by O_2 plasma oxidation. We confirmed a conduction bistability of ~10^2 and a threshold voltage for a set state of 3 V. Al nanocrystals surrounded by amorphous Al_2O_3 were formed in the α-NPD. They presented seven different reversible current paths for an electron charge or discharge on the nanocrystals. The current slightly increased with an applied bias from 0 V to V_{th} (a high resistance state (I_{off})), abruptly increased with an applied bias from V_{th} to V_p, decreased with an increasing applied bias from V_p to V_e (a negative differential resistance (NDR) region), and slightly increased with an applied bias above V_e. After sweeping the first applied voltage from 0 to 10 V (erase), a second applied bias was swept from 0 to V_p (program), where the current followed a high resistance state (I_{off}). Next, a third applied bias was swept from 0 to V_p again, where the current followed a low resistance state (I_{on}). Surprisingly, the ratio of I_{on} to I_{off} was ~1×10^2, which is enough current difference to be nonvolatile memory behavior. These I-V characteristics under a positive applied bias were symmetrically repeated under a negative applied bias. All the current sweeping paths were reproducible and symmetrical for an applied bias polarity. In particular, our device demonstrated multi-level nonvolatile memory behavior. It also revealed the current conduction mechanism for each of its operation regions. We observed that the high resistance and low resistance regions followed space-charge-limited current conduction, the V_{th} to V_p and V_{NDR} to V_e regions followed precisely thermionic-field-emission current conduction, and the above V_e regions followed space-charge-limited current conduction.

INTRODUCTION

Organic material devices have a sandwich structure of top and bottom metal electrodes separated from metal nanocrystals by conductive organic material layers. Small-molecular-weight organic nonvolatile memory using AIDCN (*2-amino-4, 5-imidazoledicarbonitrile*), Alq3 (*aluminum tris(8-hydroxyquinoline)*), or α-NPD (*N,N'-bis(1-naphthyl)-1,1'biphenyl4-4''diamine*) have been reported [1-4]. They exhibit a response time of ~10 ns and a current conduction bistability (i.e., the ratio of high-current conduction to low-current conduction) of more than 1×10^2 [1, 5]. They contain Al nanocrystals, produced by Al evaporation at rates below 0.3 Å/s, in the metal layer in the middle [1, 6]. These nanocrystals are irregular and non-uniform because they are deposited under very low evaporation conditions. As a result, the reproducibility of the retention and endurance characteristics of the devices is poor. The current conduction mechanism by which they achieve small-molecular-weight organic nonvolatile memory could also not be clarified because the reported experimental results were not repeatable. Therefore, we investigated the electrical, structural, and chemical properties of small-molecular-weight organic nonvolatile memory to clarify its current conduction mechanism. We demonstrate that this mechanism is different, e.g., space charge limited current (SCLC), thermionic emission, and F-N tunneling, for each voltage region, i.e., I_{on}, I_{off}, V_{th} to V_p, V_{NDR} to V_e, and above V_e.

EXPERIMENT

We fabricated $4F^2$ crossover memory cells with a sandwich structure of a bottom Al electrode, lower conductive organic layer (α-NPD), Al nanocrystals surrounded by Al_2O_3, upper conductive organic layer (α-NPD), and top Al electrode (Fig. 1). To isolate the crossover cells from the substrate, they were fabricated on thermally grown SiO_2 on an Si wafer. All the fabrication processes were carried out in the same evaporation chamber without breaking the vacuum to avoid contamination, e.g. H, O, N, Cl, and F atoms and H_2O molecules. After cleaning the SiO_2/Si substrate, we thermally evaporated the bottom Al electrode at 5.0 Å/s under a 10^{-5}-Pa chamber pressure by using a 1-mm line width shadow mask. The first conductive organic layer (α-NPD) was thermally evaporated at 1.0 Å/s by using a 7 x 7 mm mask. Subsequently, the Al layer in the middle was thermally evaporated at 1.0 Å/s by using a 4 x 4 mm mask, and the wafer was transferred to an O_2 plasma oxidation chamber to oxidize the Al layer to make Al nanocrystals. Next, the second conductive organic layer was thermally evaporated at 1.0 Å/s by using a 7 x 7 mm mask. Finally, the top Al electrode was thermally evaporated at 5.0 Å/s by using a mask that was cross patterned against the bottom Al electrode (Fig. 1(a)). The cell size of the device is 4 mm^2.

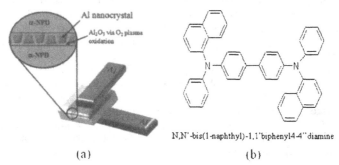

(a) (b)

Fig. 1. Small-molecular-weight organic nonvolatile memory cell with Al/α-NPD/Al nanocrystals surrounded by Al_2O_3/α-NPD/Al device structure: (a) structure of cell, (b) chemical structure of α-NPD.

After fabricating the cells, we examined their I-V characteristics at room temperature. The sample for cross-sectional transmitted electron-microscopy (X-TEM) observation was prepared using rough and fine focused-ion-beam etching. The crystal structure of the Al nanocrystals was characterized using high-voltage electron microscopy (1.25 MV).

DISCUSSION

Figure 2(a) shows the I-V characteristics of our fabricated nonvolatile memory. The first voltage was swept from 0 to 10 V. The threshold (V_{th}), program (V_p), and erase (V_e) voltages were 2.8, 4.5, and 7.2 V, respectively. The voltage between V_p and V_e was a region of negative differential resistance (NDR).

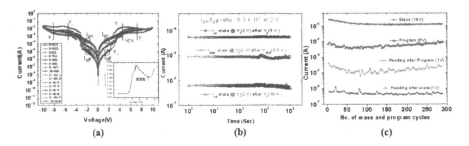

(a) (b) (c)

Fig. 2. Electrical characteristics of small-molecular-weight organic nonvolatile memory: (a) DC I-V curves (seven different reversible current paths), (b) retention test result, (c) DC endurance.

The current slightly increased with an applied bias from 0 V to V_{th} (2.8 V) (a high resistance state (I_{off})). It abruptly increased with an applied bias from V_{th} to V_p, decreased with an increasing applied bias from V_p to V_e (an (NDR) region), and increased slightly with an applied bias above V_e. After the first sweep, a second applied bias was swept from 0 to V_p, where the current followed a high resistance state (I_{off}). Next, a third applied bias was swept from 0 to V_p again, where the current followed a low resistance state (I_{on}). Surprisingly, the ratio of I_{on} to I_{off} (memory margin) was ≈1.12 x 10^2, which is a sufficient current difference to be nonvolatile memory behavior. To show a multi-level program capability, a fourth applied bias was swept from 0 to V_{NDR}, where the current followed an intermediate state (I_{inter}). Therefore, the device showed three-current-level nonvolatile memory behavior. Finally, the bias was swept from 0 to V_e, where the current followed a high resistance state (I_{off}). Figure 2(b) shows the retention characteristics of another small-molecular-weight organic nonvolatile memory. Three current (resistance) states were demonstrated: I_{on} (reading at 2 V after programming at 5 V), I_{inter} (reading at 2 V after intermediate-state programming at 6.5 V), and I_{off} (reading at 2 V after erasing at 10 V). This memory showed a memory margin of 8.3 x 10^1. The DC endurance characteristics of another memory are shown in Fig. 2(c). A memory margin of about 4.89 x 10^1 was sustained for up to about 300 erase-and-program cycles.

We investigated the crystal structure of the memory device by x-TEM (200-kV acceleration voltage) and HVEM (1.25-MV acceleration voltage). Figure 3(a) shows an x-TEM image of the device shown in Fig. 1(a), obtained with a 200-kV acceleration voltage. This image confirms that the device had three layers between the top and bottom Al electrodes and that the Al layer in the middle consisted of nanocrystals and an amorphous layer. Figure 3(b) shows an HVEM image at a 1.25-MV acceleration voltage of the section shown in Fig. 3(a). The Al nanocrystals were ~20-nm wide by ~23-nm high, uniformly distributed, and well isolated (almost 3 nm) from each other by the surrounding Al_2O_3. Their crystal structure was pure poly-aluminum with a face-centered cubic structure, and the main directions of the crystal growth were (111), (200), and (220) on the Miller index. To determine the mechanism of the memory, we investigated the temperature dependence of its DC I-V characteristics.

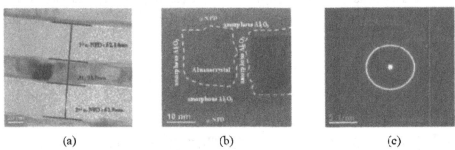

(a)   (b)   (c)

Fig. 3. Physical analysis of small-molecular-weight organic nonvolatile memory: (a) x-TEM image (at 200-kV acceleration voltage), (b) high voltage electron microscopy (HVEM) image (at 1.25-MV acceleration voltage), (c) ring pattern of Al nanocrystals.

The memory behavior disappeared at near 220 K, indicating a thermal emission of the electrons trapped on the Al nanocrystals. The regions from 0 V to V_p (I_{off}) and above V_e followed the conduction mechanism of SCLC (represented by $J = 8\varepsilon_i \mu V^2 / 9d^3$), the V_{th} to V_p and V_p to V_e (NDR) regions followed the conduction mechanism of thermionic-field-emission (represented by

$J \propto V^2 \exp(-b/V) \cdot T^2 \exp[\frac{q}{kT}(a\sqrt{V} - \phi_B)]$), and the region of reading after programming (I_{on}) followed the conduction mechanism of SCLC combined with thermionic emission (represented

by $J \propto \frac{8\varepsilon_i \mu V^2}{9d^3} \cdot T^2 \exp[\frac{q}{kT}(a\sqrt{V} - \phi_B)]$).

CONCLUSION

We fabricated small-molecular-weight organic nonvolatile memory with a sandwich structure of Al/α-NPD/Al nanocrystals surrounded Al_2O_3/α-NPD/Al. This memory showed negative differential resistance (NDR), representing multi-level program capability. The memory, with Al nano-crystals embedded in α-NPD layers, demonstrated electrical bistability, such as a I_{on}/I_{off} ratio (memory margin) of ~1.12 x 10^2, V_{th} of ~2.8 V, V_p of ~4.5 V, and V_e of ~7.2 V. Devices fabricated with in-situ O_2 plasma oxidation showed reversible current paths for erase and program. The current conduction mechanism varies depending on the operation region. These experiments contribute to the development of a small-molecular-weight organic nonvolatile memory with multi-current states in a $4F^2$ memory cell.

ACKNOWLEDGMENTS

This project was supported by "The National Research Program for 0.1-Terabit Non-volatile Memory Development" sponsored by the Korean Ministry of Commerce, Industry, and Energy.

REFERENCES

1. L. P. Ma, J. Liu, and Y. Yang, Appl. Phys. Lett. 80, 2997 (2002).
2. L. Ma, S. Pyo, J. Ouyang, Q. Xu, and Y. Yang, Appl. Phys. Lett. 82, 1419 (2003).
3. L. D. Bozano, B. W. Kean, V. R. Deline, J. R. Salem, and J. C. Scott, Appl. Phys. Lett. 84, 607 (2004).
4. J. G. Park, G. S. Lee, K. S. Chae, Y. J. Kim, and T. Miyata, Korean Phys. Soc. 48, 1505 (2006).
5. Y. Yang, J. Ouyang, L. Ma, R. J.-H. Tseng, and C.-W. Chu, Adv. Funct. Mater. 16 (2006) 1001.
6. L. D. Bozano, B. W. Kean, M. Beinhoff, K. R. Carter, P. M. Rice, and J. C. Scott, Adv. Funct. Mater. 15 (2005) 1933.

Nanoparticle-based Organic
Memory

Mater. Res. Soc. Symp. Proc. Vol. 1071 © 2008 Materials Research Society   1071-F04-04

Imaging and Elemental Analysis of Polymer/Fullerene Nanocomposite Memory Devices

Ari Laiho[1], Jayanta K. Baral[2,3], Himadri S. Majumdar[2], Daniel Tobjörk[2,3], Janne Ruokolainen[1], Ronald Österbacka[2], and Olli Ikkala[1]
[1]Department of Engineering Physics and Center for New Materials, Helsinki University of Technology, P.O. Box 5100, FIN-02015 TKK, Espoo, Finland
[2]Department of Physics and Center for Functional Materials, Åbo Akademi University, Porthansgatan 3, FIN-20500, Turku, Finland
[3]Graduate School of Materials Research, Universities of Turku, Turku, Finland

ABSTRACT

In this report we study the morphology and chemical composition of a nanocomposite memory device where the active device layer is sandwiched between two aluminum electrodes and consists of a nanocomposite of polystyrene (PS) and [6,6]-phenyl-C_{61} butyric acid methyl ester (PCBM). The morphology of the active layer is imaged both in plan-view and cross-sectional view by using transmission electron microscopy (TEM). We introduce two techniques to prepare the cross-sections for TEM, namely, a conventional technique based on microtoming and secondly nanostructural processing with focused ion beam (FIB). Based on the morphology studies we deduce that within the used concentrations the PCBM forms spherical nanoscale clusters within the continuous PS matrix. The chemical composition of the device is determined by using X-ray photoelectron spectroscopy (XPS) and it shows that the thermal evaporation of the aluminum electrodes does not lead to observable inclusion of the aluminum into the active material layer.

INTRODUCTION

The research on organic memory devices is an emergent branch in the field of organic electronics. When such a memory device is subjected to an electrical bias, the device can exhibit two distinct levels of conductivity that can then be used to define the "0" and "1" state of the memory bit. Another mechanism, besides the electrical bistability, that could possibly be used for memory applications [1] is the so called negative differential resistance (NDR) where within a certain voltage range, the current density decreases as the applied voltage is increased. So far, several different device compositions have been introduced [2-6] but their working mechanism still remains a topic of ongoing discussion [7]. It has been even suggested that the electrical characteristics of such sandwich structure devices can be explained by unintentional inclusion of aluminum into the intermediate polymer thin film while thermally evaporating the top aluminum electrode onto the polymer layer [8] or doping of the metals into the organics by application of electric field i.e. "electroforming" [9]. Moreover, Cölle et al. emphasized the role of the electrodes for obtaining reversible switching and showed that the organic material in between the sandwiched device had only minor influence [10]. They concluded that the resistive switching could be due to the breakdown of a native oxide layer at the aluminum electrode interface and transport through filaments.

There are a limited number of publications where the device morphology of an organic memory device has been investigated together with the device performance [2]. In a recent article [11], we have shown the feasibility of transmission electron microscopy (TEM) and energy dispersive X-ray spectroscopy (EDX) in analysis of the device cross-sections. The results suggested that thermal evaporation of the aluminum electrodes does not lead to the inclusion of aluminum into the active material medium.

Here we expand this research and study the device cross-sections in more detail and show that focused ion beam (FIB) nanostructural processing is a feasible technique for preparing the cross-sections. In addition we use X-ray photoelectron spectroscopy (XPS) to determine the chemical composition of the memory devices as a function of depth and conclude that there is hardly any aluminum within the active device layer due to thermal evaporation of the aluminum electrodes. Working mechanism of the reported memory devices will be discussed in detail in conjunction with the morphology elsewhere [12] whereas in this communication we concentrate more on the imaging and elemental analysis techniques.

EXPERIMENTAL DETAILS

Polystyrene (PS, 50:50 mixture of $M_w = 4000$ g/mol and $M_w = 200000$ g/mol) was obtained from Aldrich and [6,6]-phenyl-C_{61} butyric acid methyl ester (PCBM) from Nano-C Inc. and they were used as received. PS and PCBM were separately dissolved in analytical grade chloroform and thereafter mixed together. We call it p wt% PCBM/PS composition when $p = 100\% \cdot m_{PCBM} / (m_{PCBM} + m_{PS})$, where m_{PCBM} and m_{PS} are the masses of PCBM and PS, respectively.

For transmission electron microscopy (TEM) the PCBM/PS chloroform solutions were spin coated onto NaCl substrates (Sigma-Aldrich, IR crystal window). For plan-view TEM, the thin films were floated off from the substrate onto de-ionized water and collected onto TEM grids. For cross-sectional TEM (for details see [11]), a thin layer of carbon was evaporated on top and bottom of the thin film. Thereafter the thin film was embedded into epoxy followed by sectioning of thin layers (70 nm thick) from the block with an ultra-microtome.

In addition to the conventional method described above, focused ion beam (FIB) nanostructural processing was also used to obtain cross-sections of the thin films. The preparation was carried out on a Helios DualBeam using an in-situ lift out method [13].

Bright-field TEM was performed on a FEI Tecnai 12 transmission electron microscope with LaB$_6$ filament operating at an accelerating voltage of 120 kV.

The devices were made from similar thin films using the so-called sandwich type structure, in which Al-PCBM/PS-Al thin films were fabricated onto thoroughly cleaned glass substrates [11]. The thicknesses of the PCBM/PS thin films were determined with a Park Scientific Instruments AUTOPROBE cp AFM and charge extraction in a linearly increasing voltage (CELIV) technique [14]. Electrical characterization of the devices was performed by using a Keithley 2400 programmable voltage source meter and a Keithley 617 programmable Electrometer or an Agilent 4142B Modular DC Source/monitor. All device fabrication and electrical characterization was carried out inside a nitrogen filled glove box.

X-ray photoelectron spectroscopy (XPS) was performed on a Physical Electronics Quantum 2000 Scanning ESCA microprobe with an Al-K$_\alpha$ X-ray source. The sample was sputtered with argon ions (4 kV) 140 times for 12 s and after each sputtering period the XPS spectrum was detected for the C1s, O1s, Al2p and Si2p peaks (from an area of 100 μm × 100 μm). The chemical composition was determined from the area of the peaks in the XPS spectrum and the composition was plotted as a function of sputtering time.

RESULTS AND DISCUSSION

In our previous work [11], we studied the morphology of PCBM/PS films at low weight percentages (<10 wt%) of PCBM in PS and reported well controlled morphology without formation of aggregates. Here we use much higher PCBM concentrations and observe significantly different morphology. A TEM micrograph from such a 40 wt% PCBM/PS thin film is shown in figure 1a where PCBM is observed to aggregate into nanoscale clusters which are surrounded by the PS matrix. Study of the plan-view TEM micrographs alone can not provide detailed information on the 3D arrangement of the clusters since TEM lacks depth sensitivity [15]. As a complementary technique to obtain information on the arrangement of the clusters in the vertical direction, we have prepared and imaged cross-sections of the thin films (see figure 1b).

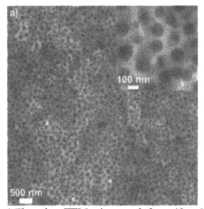

Figure 1. a) Plan-view TEM micrograph from 40 wt% PCBM/PS thin film showing the spherical PCBM clusters within the continuous PS matrix. b) TEM micrograph from a cross-section of 40 wt% PCBM/PS thin film that has been prepared using the conventional method. The wavy black line is the evaporated carbon layer that has been prepared to avoid penetration of epoxy into the thin film and to observe the interfaces of the thin film better. The scale bar in b) is 200 nm.

The preparation of the cross-sections using our conventional method is done by embedding the thin films into epoxy and thereafter cutting thin sections from the epoxy block with a microtome (for details see [11]). To avoid penetration of epoxy into the thin film, thin layers of carbon are evaporated to both interfaces of the thin film prior to embedding into epoxy. Use of this technique requires that the contact between the epoxy and the carbon layers is sufficiently strong so that the sectioning does not lead to film rupture. FIB processing was used as another method to prepare cross-sections from the thin films and a TEM micrograph from such a FIB processed cross-section is shown in figure 2. The figure confirms the observation that the PCBM aggregation leads to spherical clusters which modify the air/film interface to become more corrugated.

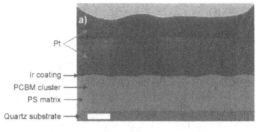

Figure 2. a) TEM micrograph from a FIB prepared cross-section of 40 wt% PCBM/PS thin film that has been spin coated onto a quartz substrate. Prior to FIB preparation the thin film has been coated with a thin Ir coating and a localized SEM- and FIB-induced Pt coating. b) TEM micrograph from the same FIB prepared cross-section with a different contrast to show the spherical PCBM clusters better. Both scale bars are 200 nm.

XPS was used to determine the chemical composition of the memory devices as a function of depth. As the sputtering process proceeds, the first contribution in XPS comes from the top aluminum electrode and the last from the glass substrate. The depth profile in figure 1a shows that the top aluminum electrode is only oxidized close to the surface since the oxygen signal starts to decay well before the aluminum signal.

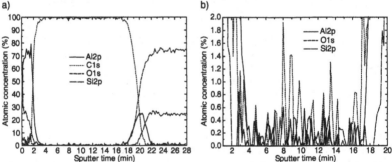

Figure 3. a) The atomic concentration of aluminum, carbon, oxygen and silicon as determined from the XPS spectra are plotted as a function of sputtering time (Cf. 0.3 nm/s sputtering speed on a silicon substrate). b) A zoom into the low atomic concentration region of figure a) is shown.

The aluminum concentration then decreases fast and merely carbon is detected when the sputtering proceeds through the PCBM/PS layer. The signal of the other elements is very low (similar concentrations for Al and Si), as can be seen in figure 1b, and are mainly due to the background noise. Aluminum content inside the polymer layer can therefore be excluded, but it should be kept in mind that the detection limit of XPS is 0,1-1 at.%. There are also some unwanted effects from the argon sputtering; a rough surface is formed and some elements might be pushed further into the sample. This explains the much broader shape (lower concentration and thicker) of the bottom aluminum electrode and that silicon from the glass substrate is detected so early. This effect can partially also be due to a small piece of the analyzed area being outside of the bottom contact. Interesting to notice is also that the oxygen content increases at the same time as the aluminum signal, which suggests that the bottom electrode is also oxidized.

The electrical behavior of the 40 wt% device follows the similar path as reported in our recent article [11]. At a threshold voltage of approximately $V_{th} = 3$ V the current density suddenly jumps two orders of magnitude which is followed by NDR. Because the initial jump at V_{th} only occurs for the first scan and the device remains at the high conductivity region thereafter, such devices could have non-volatile application. Moreover, the current density continues to follow the same increasing trend with increasing PCBM concentration as reported in [11].

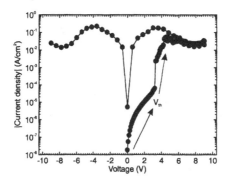

Figure 4. Absolute current density as a function of voltage for 40 wt% PCBM/PS device (275 nm thick).

CONCLUSIONS

In conclusion, the PCBM molecules were shown to aggregate into nanoscale spherical clusters within the polystyrene matrix. FIB processing was shown to be a versatile technique to prepare cross-sections even from the soft nanocomposite materials described here. The elemental analysis of the devices showed that thermal evaporation of the aluminum electrodes does not lead to observable inclusion of aluminum into the active PCBM/PS layer. The electrical characterization of such devices showed existence of an initial threshold voltage and NDR. We believe that the reported techniques for studying the device morphology and chemical composition of the memory devices will become important in understanding the device behavior.

133

ACKNOWLEDGMENTS

The authors are particularly grateful to Dr. Steve Reyntjens from the FEI Company for FIB preparation of the cross-sections for TEM. Dr. Henrik Sandberg, Marja Vilkman and Dr. Robin H. A. Ras are acknowledged for fruitful scientific discussions.

REFERENCES

1. Y. Nakasha, and Y. Watanabe, U.S. Patent No. 5 390 145 (14 Feb 1995).
2. L. D. Bozano, B. W. Kean, M. Beinhoff, K. R. Carter, P. M. Rice, and J. C. Scott, *Adv. Funct. Mater.* **15**, 1933 (2005).
3. H. S. Majumdar, J. K. Baral, R. Österbacka, O. Ikkala, and H. Stubb, *Org. Electron.* **6**, 188 (2005).
4. J. Y. Ouyang, C. W. Chu, C. R. Szmanda, L. P. Ma, and Y. Yang, *Nat. Mater.* **3**, 918 (2004).
5. S. Paul, A. Kanwal, and M. Chhowalla, *Nanotechnology* **17**, 145 (2006).
6. Y. Yang, J. Ouyang, L. P. Ma, R. J. H. Tseng, and C. W. Chu, *Adv. Funct. Mater.* **16**, 1001 (2006).
7. J. C. Scott, and L. D. Bozano, *Adv. Mater.* **19**, 1452 (2007).
8. W. Tang, H. Z. Shi, G. Xu, B. S. Ong, Z. D. Popovic, J. C. Deng, J. Zhao, and G. H. Rao, *Adv. Mater.* **17**, 2307 (2005).
9. K. Efimenko, V. Rybka, V. Svorcik, and V. Hnatowicz, *Appl. Phys. A* **67**, 503 (1998).
10. M. Cölle, M. Büchel, and D. M. de Leeuw, *Org. Electron.* **7**, 305 (2006).
11. J. K. Baral, H. S. Majumdar, A. Laiho, H. Jiang, E. I. Kauppinen, R. H. A. Ras, J. Ruokolainen, O. Ikkala, and R. Österbacka, *Nanotechnology* **19**, 035203 (2007).
12. A. Laiho, J. K. Baral, H. S. Majumdar, F. Jansson, A. Soininen, R. Österbacka, and O. Ikkala, (submitted).
13. S. Reyntjens, and R. Puers, *J. Micromech. Microeng.* **11**, 287 (2001).
14. G. Juška, K. Arlauskas, M. Viliūnas, K. Genevičius, R. Österbacka, and H. Stubb, *Phys. Rev. B* **62**, R16235 (2000).
15. D. B. Williams, and C. B. Carter, *Transmission electron microscopy: a textbook for materials science*, (Plenum Press, New York, 1996) p. 10.

Mater. Res. Soc. Symp. Proc. Vol. 1071 © 2008 Materials Research Society   1071-F05-12

Small Molecular Organic Nonvolatile Memory Fabricated with Ni Nanocrystals Embedded in Alq3

Younghwan Oh[1], Woosik Nam[1], Gonsub Lee[1], Jeagun Park[1], and Yongbok Lee[2]
[1]Department of Electrical & Computer Engineering, Hanyang university, Tera-bit Nonvolatile Memory Development Center, room #101, HIT, 17 Haengdang-dong, Seongdong-gu, Seoul, 133-791, Korea, Republic of
[2]Electron Microscopy Team, Korea Basic Science Institute, Deajeon, 305-333, Korea, Republic of

ABSTRACT

Recently, organic nonvolatile memory has attracted much interest as a candidate device for next generation nonvolatile memory because of its simple process, small device area, and high speed. To investigate electrical characteristics of small molecular organic nonvolatile memory with Ni as a middle metal layer, we developed a small molecular organic nonvolatile memory with the device structure of Aluminum tris (8-hydroxyquinolate) (Al/Alq3), Ni nanocrystals, and Alq3/Al. A high vacuum thermal deposition method was used for the device fabrication. It is critical that the fabrication process condition for Ni nanocrystals be optimized, including ~100 Å thickness, 0.1 Å/sec-evaporation rate, and in-situ plasma oxidation for effective oxidation. The reasons we chose Ni for the middle metal layer are that Ni has a smaller grain boundary, which is beneficial for scaling down and has a larger work function (~5.15 eV) that can make a deep quantum well in an energy band diagram, compared with that of Al. Our device showed an electrical nonvolatile memory behavior including V_{th} of ~2 V, V_w (write) of ~3.5 V, negative differential region (NDR) of 3.5~7 V, V_e (erase) of 8 V, and symmetrical electrical behavior at reverse bias. In addition, an interesting behavior of electrical properties was that, although retention and endurance characteristics were similar to the Al device, the I_{on}/I_{off} ratio was greater than 10^4 at V_r (read) of 1 V. This value of the Ni device was higher than 10^2 compared to that of the Al device. Also, small molecular organic nonvolatile memory with a Ni middle layer with α-NPD at same fabrication condition showed more unstable characteristics than Alq3. We can speculate that there is a relationship in fabrication condition between the middle metal material and the organic material. Finally, we conclude that our device with a Ni nanocrystals middle layer is more reliable and useful for small molecular organic nonvolatile memory.

INTRODUCTION

Next generation nonvolatile memory is studied actively in many industrial and academic labs. Yang et al. (UCLA) reported on a high performance organic bistable device cell of a sandwich structure, which had Al nanocrystals embedded in a small molecular organic material 2-amino-4, 5-imidazoledicarbonitrile (AIDCN) between Al electrodes [1-3]. They fabricated Al nanocrystals by using a very low evaporation rate. Scott et al. (IBM) also reported on a device by using small molecular organic Alq3 (Aluminum tris (8-hydroxyquinoline)), which had the same structure of the UCLA's group [4]. However, the electrical I-V characteristics of the IBM device are different from those of the UCLA device. The electrical I-V characteristics of the IBM device showed negative differential region (NDR) and an intermediate state by using NDR, which showed that it is possible to adjust the resistance state and to realize multi level cell (MLC) devices. However these devices are low reproducibility and show a small I_{on}/I_{off} memory margin ($\approx 10^2 \sim 10^3$) [4]. Therefore, we proposed a new nanocrystal material, Ni nanocrystals, and an effective fabrication process using O_2 plasma oxidation. The device for inserting Ni nanocrystals showed large memory margin and reproducibility. In addition, Ni nanocrystals have small grain boundaries that are suitable for integration circuits. We then tested the devices' nonvolatile memory characteristics such as memory margin, data retention time, and endurance of erase and program cycles.

EXPERIMENT

Our device memory cells were fabricated with a crossover sandwich structure consisting of a bottom Al electrode, lower conductive organic layer (Alq_3), Ni nanocrystals surrounded by NiO, upper conductive organic layer (Alq_3), and a top Al electrode, as shown in Figure 1. To isolate them from the substrate, the crossover cells were fabricated on thermally grown 700-nm SiO_2 on a Si wafer. All fabrication processes were carried out in the same evaporator without breaking the vacuum to prevent contamination (e.g. H, O, N, Cl, and F atoms). The 80-nm-thick bottom Al electrode was thermally evaporated at a rate of 5 Å/s and a chamber pressure of 10^{-5} Pa using a first shadow mask with a line width of 3 mm. The 30-nm-thick lower conductive organic layer (Alq_3) was thermally evaporated at a rate of 1.0 Å/s using a 7 mm x 7 mm second mask. Next, the middle Ni layer was thermally evaporated at a rate of 0.1 Å/s using a 5.5 mm x 5.5 mm third shadow mask, and the wafer was transferred to the O_2 plasma chamber to oxidize the Ni middle layer (Ni nanocrystals surrounded by NiO). Next, the second 30 nm-thick conductive organic layer (Alq_3) was thermally evaporated at a rate of 1.0 Å/s using a 7 mm x 7 mm second mask. Finally, the 80-nm-thick top Al electrode was thermally evaporated at a rate of a 5 Å/s using a fourth mask, which was cross-patterned against the bottom Al electrode. All memory fabrication processes were carried out in an in-situ multi-chamber apparatus, which made it possible to change and align the masks. After fabrication, we first evaluated the current versus voltage (I-V) characteristics. The sample for cross sectional transmission electron microscope (TEM) observation was prepared by using focused-ion-beam (FIB) etching.

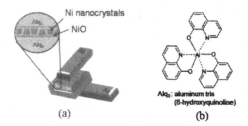

(a) (b)

Figure 1. Organic nonvolatile memory cell (cell size 3 mm x 3 mm) with an $Al/Alq_3/Ni$ nanocrystals/Alq_3/Al device structure: (a) perspective view of memory cell, (b) chemical structure of Alq_3

The crystal structure of the Ni nanocrystals was observed using ultra-high-voltage TEM (1.2 MV). And composition of the Ni middle layer was observed using an energy dispersive spectroscopy (EDS) profile.

DISCUSSION

Electrical I-V characteristics

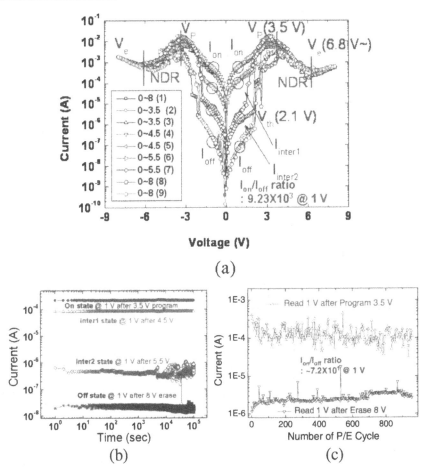

Figure 2. Electrical characteristics of organic nonvolatile memory cell: (a) I-V curves, (b) retention, and (c) endurance.

Figure 2(a) shows the I-V characteristics of our memory cells. The sandwich device structure caused the symmetrical I-V characteristics. First, the current increased slightly at applied biases from 0 V up to V_{th} ($\approx$2.1 V). Second, the current increased abruptly at bias region from V_{th} to V_p (program voltage: $\approx$3.5 V). Third, it decreased on the contrary at $V_p \sim V_e$ ($\approx$6.8 V~). This current region is called the NDR. Finally, the current increased again at applied biases above V_e (6.8 V up to 8 V). After the first applied voltage was swept from 0 to 8 V (called erase), as shown in figure 2(a), the second applied bias was swept from 0 to V_p (called program), and then the current followed the high-resistance state (I_{off}). After the second applied bias (3.5 V: program voltage) was swept, the third applied bias (reading after program) was again swept from 0 to V_p, and then the current followed I_{on}. This shows that the device reads I_{on} (the high current state) at 1 V after erase (biasing 3.5 V or V_p), otherwise the device reads I_{off} (the low current state) at 1 V after program (biasing above V_e). The ratio of I_{on} to I_{off} (memory margin) was $\approx$9.23 x 10^3, which is a sufficient current difference (bistable resistance difference) for nonvolatile memory. After the third applied bias (3.5 V: reading after program) was swept, the fourth applied bias was swept from 0 to one bias point in NDR ($V_p \sim V_e$), and then the current followed I_{inter}. As one bias point closer to V_e, I_{inter} occupied a smaller value, as shown in figure 2(a), which demonstrates that developing multilevel nonvolatile memory is possible. Figure 2(b) shows the retention characteristic of another memory cell. Four current states (resistance states) were demonstrated: I_{on} (reading at 1 V after 3.5 V programming), I_{int1}. (reading at 1 V after intermediate-state programming at 4.5 V), I_{int2}. (reading at 1 V after intermediate-state programming at 5.5 V), and I_{off} (reading at 1 V after erasing at 8 V). This memory cell showed a superior memory margin (I_{on}/I_{off} ratio) of 9.1x10^3 and charging capability of 10^5 secs. The DC endurance characteristics of another memory cell are shown in figure 2(c). A memory margin of about 7.2 x 10^1 was sustained for up to about 10^3 erase and program cycles. These improvements of memory characteristics are from the inserted Ni nanocrystals. Ni has a large work function (5.15 eV), and it shows a deep quantum well in the energy band diagram.

Cross sectional TEM analysis

A transmission electron microscope (TEM) image of one of the memory cells we fabricated is shown in figure 3(a). The thicknesses of the upper Alq3 layer, the middle layer, and the lower Alq3 layer were respectively about 29.2 nm, 10.2 nm, and 32.1 nm. An energy dispersive spectroscopy (EDS) profile is shown in Figure 3(b). In this profile, we could see a peak of oxygen and Ni in the middle Ni layer, which demonstrates that the Ni middle layer consists of partially oxidized Ni. A magnified TEM image (obtained at a 1.2-MV acceleration voltage) of the cell shown in figure 3(a) is shown in figure 3(c). The fringe of Ni is shown in figure 3(c) and many fringes in other directions are overlaid. Figure 3(d) is a diffraction ring pattern analysis from one fringe of Ni in figure 3(c). The crystal structure of Ni was face-centered cubic of pure poly-nickel, predominantly composed of (111), (200), and (220). These results infer that the middle Ni layer is composed of Ni nanocrystals that are less than 10 nm in size (Ni middle layer thickness is below 10 nm). On the other hand, our organic nonvolatile memory cell samples with Al nanocrystals usually showed memory characteristics at middle layer thickness of above 20 nm and a TEM image demonstrated that the Al fringe size was about 20 nm.

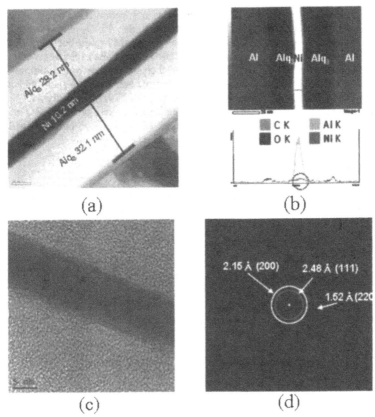

(a)   (b)

(c)   (d)

Figure 3. Physical and chemical analysis of an organic nonvolatile memory cell (cell size 3 mm x 3 mm): (a) TEM image (200-kV acceleration voltage), (b) EDS profile, (c) magnified TEM image (1.25-MV acceleration voltage), and (d) ring pattern analysis.

Therefore, Ni has the advantage of scalability, and organic nonvolatile memory cells fabricated with an organic conductive layer embedded with uniform roughly 10-nm Ni nanocrystals could provide tera-bit-level memory density.

CONCLUSIONS

Until now Al nanocrystals were usually used in sandwich structures of small molecular organic memory. This type of device cell showed rather small I_{on}/I_{off} memory margins ($\approx 10^2 \sim 10^3$) for multilevel nonvolatile memory and poor reproducibility because of difficult nanocrystal fabrication (effective oxidation is very important) and rough nanocrystal uniformity [5], plus scalability problems because of a lager grain boundary. Our organic nonvolatile memory cell with Ni nanocrystals inserted in Alq_3 showed improvements to the above characteristics. First, a large work function provides a superior charging capability, which provides large I_{on}/I_{off} and reliable memory characteristics. Second, O_2 plasma oxidation allows fabrication of uniform nanocrystals possible. Finally, the small grain boundary of Ni makes application for ultra-large scale integration possible. Therefore, these experiments will contribute to the development of small molecular nonvolatile memory devices with multilevel cells and realization of a reliable and useful small molecular organic nonvolatile memory. In addition, we conducted experiments using another organic materials (N, N'-bis(1-naphthyl)-1,1'biphenyl 4-4''diamine: α-NPD) using the same structure and process. Results from these experiments showed very poor memory characteristics. Therefore, the next step in our work will to try to understand the relationship between variable metals and organic materials.

ACKNOWLEDGMENTS

This project was supported by "The National Research Program for 0.1-Terabit Non-volatile Memory Development" sponsored by the Korean Ministry of Commerce, Industry and Energy.

REFERENCES

1. L. P. Ma, J. Liu and Y. Yang, Appl. Phys. Lett. 80, 2997 (2002).
2. L. P. Ma, S. M. Pyo, J. Ouyang, Q. Xu and Y. Yang, Appl. Phys. Lett. 82, 1419 (2003).
3. J. He, L. Ma, J. Wu and Y. Yang, J. Appl. Phys. 97, 064507 (2005).
4. L. D. Bozano, B. W. Kean, V. R. Deline, J. R. Salem and J. C. Scott, Appl. Phys. Lett. 84, 607 (2004).
5. J. C. Scott, L. D. Bozano, Adv. Mater. 19, 1452 (2007)
6. J. G. Park, G. S Lee K. S. Chae, Y. J. Kim and T. Miyata, J. Korean Phys. Soc. 48, 1 (2006).

Mater. Res. Soc. Symp. Proc. Vol. 1071 © 2008 Materials Research Society   1071-F05-14

Effect of Au Nanocrystals Embedded in Conductive Polymer on Non-volatile Memory Window

Hyun Min Seung, Jong Dae Lee, Byeong-Il Han, Gon-Sub Lee, and Jea-Gun Park
Division of Nanoscale Semiconductor Engineering, Hanyang University, Nano SOI process
Laboratory, Room #101,HIT, 17 Haedang-dong, Seoungdong-gu, Seoul,
133-791, Korea, Republic of

ABSTRACT

Nonvolatile memory devices based on the bistability phenomenon have been researched for the next generation of nonvolatile memory, because of their simple structure and easy fabrication process. Bistability was observed in a poly(n-vinylcarbazole) (PVK) layer, that contain many small, discrete Au nanocrystals, and was sandwiched between Al electrodes. The effect of the Au nanocystals embedded in PVK was investigated and shown to induce a different type of bistability from that exhibited by devices with other structures. The results suggest that Au nanocrystals are essential for conductive polymer memory device to have stable nonvolatile memory behavior and window.

INTRODUCTION

Recently, many researchers have investigated polymer nonvolatile memory devices because of their low-cost, flexible, simple fabrication [1-4]. The memory phenomenon in these devices is based on the electrical bistability of the material, which has two resistance states that can be set to two different voltages [5-11]. The Simmons-Verderber (SV) model has been used to explain this change in conductivity [12]. Filament formation and destruction [13,14] and electro-reduction [15-17] have also previously been considered. Nonvolatile memory devices that use discrete nanocrystals as charge storage sites and exhibit bistability have also been reported [10]. We have fabricated polymer memory devices with structures consisting of Al / PVK / Al, Al(with O_2 plasma treatment) / PVK / Al, and Al / PVK / Au nanocrystals / PVK / Al. Depending on the structure, these devices exhibit different nonvolatile memory behavior. In this letter, we discuss how the nonvolatile memory behavior with Au nanocrystals embedded in PVK differ from that of structures without nanocrystals.

EXPERIMENT

We fabricated three devices with different structures between the Al top and bottom electrodes. The devices were fabricated on cleaned SiO_2. The top and bottom electrodes were deposited on the substrate by thermal evaporation in a vacuum chamber (pressure ~ 10^{-6} torr) The first type of device had only a polymer layer between the electrodes. The polymer (PVK) was dissolved with chloroform, spin-coated on the bottom Al electrode, and baked at 120°C for 2 min to evaporate the solvent away. The second type of device was fabricated by the same processes used for the first type, with the addition of O_2 plasma oxidation after the Al bottom electrode was deposited. For the third type of device, a 5-nm-thick Au film was deposited on the spin-coated PVK layer after the bottom electrode deposition. Additional PVK was then spin-coated on the Au film and baked. Next, the device was cured at 300°C for 2 h in air to produce

Au nanocrystals. Finally, the Al for the top electrode was deposited on the cured device. Figure 1 shows a cross section of the device structure for the third type and the chemical structure of PVK. The electrical characteristics of the devices were measured with an Agilent 4270B

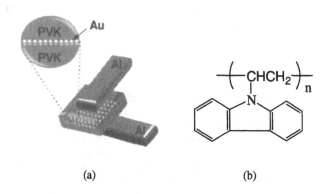

(a)   (b)

Figure 1. (a) Device structure consisting of Al / PVK / Au nanocrystals / PVK / Al , (b) The Chemical structure of poly(N-vinylcarbazole) (PVK).

DISCUSSION

We measured the electrical characteristics of the devices for bistability at room temperature in air. Figure 2 shows the typical electrical characteristics obtained by sweeping the voltage from 0 to 8 V. The first type of device showed ohmic electrical behavior, indicating bistability not based on the polymer characteristics. The second and third device types exhibited different bistable behaviors. The second type showed a low -to -high current path at increased positive bias and a high -to -low current path at increased negative bias. This indicates I-V behavior similar to that of resistive memory [16-17]. The ON and OFF states could be set at voltages close to 8 V respectively and could be read at 1 V. After the device was programmed by sweeping the voltage from 0 to 8 V, the current followed the high -current curve and stayed in the ON state until applying a voltage near -8 V. The third device type exhibited a low -to -high current path not only at increased positive bias but also at increased negative bias. When the voltage was increased from zero in the OFF state, the current increased rapidly at the threshold voltage (V_{th}) and exhibited a regime of negative differential resistance (NDR) [21-30]. Furthermore, the ON and OFF states could be set at voltages close to Vprogram (or V_p) and Verase (or V_e), respectively, and could be read at 1 V. After the device was programmed by sweeping the voltage from 0 to V_p, the current followed the upper curve and stayed in the ON state at applied voltages between 0 V and V_p. The values of V_{th}, V_p, and V_e were ~2.5, ~3.7, and ~6 V, respectively. In particular, this device exhibited nonvolatile memory behavior, with bistability (I_{ON}/I_{OFF}) of more than 1×10^3. The third type of device demonstrated the electrical behavior reported for nanocrystals embedded in polymer nonvolatile memory devices.

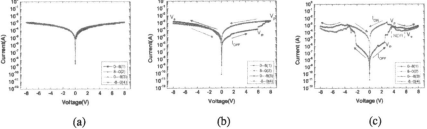

(a)   (b)   (c)

Figure 2. Typical electrical characteristics obtained by sweeping the voltage from 0 to 8 V.: (a) first device type , showing ohmic I-V behavior, (b) second type and (c) third type. The second and third device types exhibited different nonvolatile memory behavior. The second device type showed behavior similar to that of resistive memory.

We found that O_2 plasma oxidation after Al bottom electrode deposition induced different nonvolatile memory behavior from that of the Au nanocrystals embedded in a conductive polymer. This suggests that the nonvolatile memory behavior of each device resulted from a different mechanism. For the second device type, the O_2 plasma treatment might have induced an interfacial change of the Al bottom electrode after oxidation, generating nonvolatile memory behavior. In a subsequent letter, we will examine the bistability mechanisms of our device in detail. The mechanism inducing bistability in the third device type consisted of charge trapping and space-charge field inhibition of injection. According to the SV model, the trapping sites are deep-level Au atoms diffused from the Al electrode, whereas those in our device were the Au nanocrystals produced by the curing process. The reported polymer-based nonvolatile memories without nanocrystals show electrical behavior similar to that of resistive memory, like the second device. Figure 3 shows the endurance characteristics of devices. The device subjected to O_2 plasma treatment showed unstable endurance characteristics, whereas those of the Au nanocrystals embedded in the conductive polymer layer were more stable and had a larger memory window, as noted above. But there was degradation of conduction current. The Device with Au nanocrystals need more optimization.

143

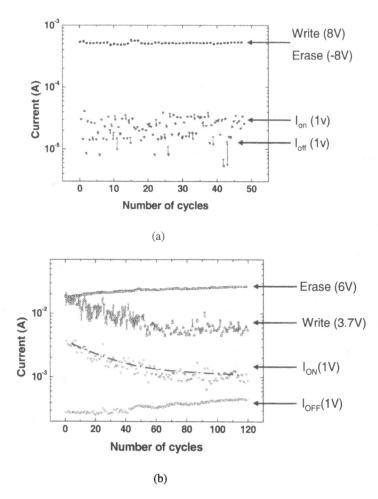

Figure 3. Endurance characteristics of (a) the second device type, and (b) the third type. The second type exhibited very unstable endurance characteristics.

CONCLUSIONS

Recently, many types of nonvolatile memory devices based on conductive polymers have been reported. One type has nanocrystals embedded in the polymer layer, while other types lack such crystals. In this letter, we have reported the effect on the memory window of embedding Au nanocrystals in the conductive polymer. We observed different nonvolatile memory behaviors in different devices, depending on the presence of Au nanocrystals. The results suggest that nonvolatile memory without Au nanocrystals operates by a mechanism similar to that of resistive memory.[16-17] In addition, the device with nanocrystals showed more stable electrical characteristics and a larger memory window than did the devices without them. The device with Au nanocrystals reported here is easy to fabricate and suggests the promise of next-generation nonvolatile memory devices.

ACKNOWLEDGMENTS

This research was supported by the Korea Ministry of Commerce, Industry and Energy, through the National Development Program for 0.1-Terabit Non-volatile Memory.

REFERENCES

1. C. W. Tang and S. A. Van Slyke, Appl. Phys. Lett. **51**, 913 (1987).
2. J. H. Burroughs, D. D. C. Bradley, A. R. Brown, R. N. Marks, K. Mackay, R. H. Friend, P. L. Bums, and A. B. Holmes, Nature (London) **347**, 539 (1990).
3. F. Garnier, R. Hajlaoui, A. Yassar, and P. Srivastava, Science **265**, 1684 (1994).
4. N. Tessler, G. J. Denton, and R. H. Friend, Nature (London) **382**, 695 (1996).
5. L. P. Ma, J. Liu, S. Pyo, and Y. Yang, Appl. Phys. Lett. **80**, 362 (2002).
6. L. P. Ma, J. Liu, and Y. Yang, Appl. Phys. Lett. **80**, 2997 (2002)
7. L. P. Ma, S. M. Pyo, J. Ouyang, Q. F. Xu, and Y. Yang, Appl. Phys. Lett. **82**, 1419 (2003).
8. L. P. Ma, J. Liu, S. M. Pyo, Q. F. Xu, and Y. Yang, Molec. Cryst. Liq. Cryst. **378**, 185 (2002).
9. J. He, L. P. Ma, J. Wu, and Y. Yang, J. Appl. Phys. **97**, 64507 (2005).
10. L. D. Bozano, B. W. Kean, M. Beinhoff, K. R. Carter, P. M. Rice, and J. C. Scott, Adv. Funct. Mater. **15**, 1933 (2005).
11. L. D. Bozano, B. W. Kean, V. R. Deline, J. R. Salem, and J. C. Scott, Appl. Phys. Lett. **84**, 607 (2004).
12. J. G. Simmons and R. P. Verderber, Proc. R. Soc. London, Ser. A **301**, 77 (1967).
13. Y. S. Lai, C. H. Tu, D. L. Kwong, and J. S. Chen, Appl. Phys. Lett. **87**, 122101 (2005).
14. G. Dearnaley, A. M. Stoneham, and D. V. Morgan, Rep. Prog. Phys. **33**, 1129 (1970).
15. A. Bandyopadhyay and A. J. Pal, J. Phys. Chem. B **109**, 6084 (2005).
16. A. Sawa, T. Fujii, M. Kawasaki, and Y. Tokurad, Appl. Phys. Lett. **85**, 4073 (2004).
17. A. Beck, J. G. Bednorz, Ch. Gerber, C. Rossel, and D. Widmer, Phys. Lett. 77, 139 (2000).
18. D. C. Kim, M. J. Lee, S. E. Ahn, S. Seo, J. C. Park, I. K. Yoo, I. G. Beak, H. J. Kim, J. E. Lee, S. O. Park, H. S. Kim, U-I. Chung, J. T. Moon, and B. I. Ryu, Appl. Phys. Lett 88, 232106 (2006).
19. M. D. Leea, C. K. Lob, T. Y. Penga, S. Y. Chena, and Y.D. Yaoa, J. Magnetism and Magnetic Materials 310, e1031 (2007).
20. D. S. Shang, L. D. Chen, Q. Wang, Z. H. Wu , W. Q. Zhang, and X. M. Li, J. Phys. D: Appl. Phys. 40, 5373–5376 (2007).
21. Q. Ling, Y. Song, S. J. Ding, C. Zhu, D. S. H. Chan, D.-L. Kwong, E.-T. Kang, and K. G. Neoh, Adv. Mater. (Weinheim, Ger.) 17, 455 (2005).

22. C. W. Chu, J. Quyang, J. H. Tseng, and Y. Yang, Adv. Mater. (Weinheim, Ger.) **17**, 1440 (2005).
23. S. Smith and S. R. Rorrest, Appl. Phys. Lett. **84**, 5019 (2004).
24. J. Chen and D. Ma, Appl. Phys. Lett. **87**, 023505 (2005).
25. W. Tang, H. Z. Shi, G. Xu, B. S. Ong, Z. D. Popovic, J. C. Deng, J. Zhao, and G. H. Rao, Adv. Mater. (Weinheim, Ger.) **17**, 2307 (2005).
26. W. J. Yoon, S. Y. Chung, P. R. Berger, and S. M. Asar, Appl. Phys. Lett. **87**, 203506 (2005).
27. J. Chen, W. Wang, M. A. Reed, A. M. Rawlett, D. W. Price, and J. M. Tour, Appl. Phys. Lett. **77**, 1224 (2000).
28. M. A. Reed, J. Chen, A. M. Rawlett, D. W. Price, and J. M. Tour, Appl. Phys. Lett. **78**, 3735 (2001).
29. J. D. Le, Y. He, T. R. Hoye, C. C. Mead, and R. A. Kiehl, Appl. Phys. Lett. **83**, 5518 (2003).
30. S. I. Khondaker, Z. Yao, L. Cheng, J. C. Henderson, Y. X. Yao, and J. M. Tour, Appl. Phys. Lett. **85**, 645 (2004).

Mater. Res. Soc. Symp. Proc. Vol. 1071 © 2008 Materials Research Society   1071-F05-21

Process Optimization of Ni Nanocrystals Formation Using O2 Plasma Oxidation to Fabricate Low-molecular Organic Nonvolatile Memory

Woo Sik Nam, Gon-Sub Lee, Sung Ho Seo, and Jea Gun Park
Electronic engineering, National Program Center for Terabit-level Nonvolatile Memory
Development, HIT 101 Handangdong Seongdonggu Hanyang University, Seoul, Korea, Seoul,
Korea, Republic of

ABSTRACT

We fabricated organic nonvolatile memory with a device structure of Al/Alq3 (*aluminum tris (8-hydroxyquinoline)*)/Ni nanocrystals surrounded by NiO/Alq3/Al. We obtained the best bistable switching characteristics at a 30-nm Alq3 thickness, 0.1-Å/sec evaporation rate, and 10-nm Ni nanocrystal layer thickness. The electrical behavior of the bistable switching devices was obtained by sweeping the voltage from 0 to 10 V. Our devices showed excellent bistable memory characteristics, such as a V_{th} of 2 V, V_p of 3 V, V_e of 5 V, and I_{on}/I_{off} ratio of greater than 10^4. We found that a region of negative differential resistance exists between V_p and V_e.

INTRODUCTION

Various kinds of organic nonvolatile memory have been reported. They can be generally classified as organic nonvolatile memory embedded with metal nanocrystals in a conductive organic layer (low-molecular organic, polymer etc.) [1-6], with a single charge-complex layer [5-8], and with a single conductive organic layer, etc. [5-6, 9-10]. We previously researched low-molecular organic nonvolatile memory with metal nanocrystals embedded in the low-molecular organic layer. This memory achieved a I_{on}/I_{off} ratio of more than 10^1. However, the nanocrystals in these devices had irregular shapes and non-uniform distribution because the nanocrystal layer contained large-grained Al and was deposited at very low evaporation rates. Accordingly, the devices showed very bad reproducible I-V characteristics. To overcome this weakness, we formed a Ni nanocrystal layer having regular shaped and uniformly distributed grains using in-situ O2 plasma oxidation. We found that the Ni nanocrystals surrounded by NiO reveal the I-V and memory characteristics of organic nonvolatile memory during fabrication. As a result, we determined the optimal conditions for forming Ni nanocrystals to obtain good bistable memory characteristics in organic nonvolatile memory embedded with Ni nanocrystals.

EXPERIMENTAL DETAILS

First, Si/SiO2 substrates were coated with Al by thermal evaporation to be the bottom electrode, which was patterned by a shadow mask technique. The bottom organic layer and the metal layer in the middle were sequentially deposited on the bottom electrode. Ni was evaporated, followed by O2 plasma oxidation for 300 sec at 200 W in a 10^{-4}-Pa vacuum. Next, the top organic layer and the top electrode were deposited. All the fabrication processes for the organic nonvolatile memories were done in an in-situ multi-chamber apparatus by simply changing and aligning the masks. In a repeat of the process, we confirmed that the amount of oxidation and Ni nanocrystals formed was in accordance with the various Ni evaporation rates, e.g., 0.1, 0.5, and 1.0 Å/sec, and with whether or not O_2 plasma oxidation was used. The I-V characteristics were measured using an Agilent 5270B, and the chemical composition of the Ni nanocrystal layer was characterized by Auger electron spectroscopy with a 10-nm beam diameter and X-ray photoelectron spectroscopy with a 0.1 x 0.4-mm beam diameter. The samples for a cross-sectional transmission electron microscope (TEM) were prepared by rough and fine focused ion beam etching. The crystal structure of the Ni nanocrystals was analyzed by ultra high voltage-TEM at a 1.2-MV acceleration voltage.

DISCUSSION

We prepared five samples under the conditions shown in Table 1. The Ni thickness was fixed at 10 nm, and various Ni evaporation rates, i.e., 0.1, 0.5, 1.0 Å/sec, were used. The use of O_2 plasma oxidation varied.

Table 1. Experimental conditions

Sample	Ni evaporation rate (Å/s)	Ni thickness (nm)	O_2 plasma oxidation
#1	1.0	10	X
#2	0.1	10	X
#3	1.0	10	O
#4	0.5	10	O
#5	0.1	10	O

Auger electron spectroscopy (AES) analysis

The results of the AES analysis are shown in Fig. 1. Figures 1(a), (b), (c), (d), and (e) are the results of samples 1, 2, 3, 4, and 5 (conditions shown in Table 1), respectively.

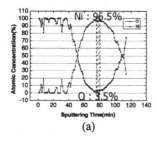
(a)

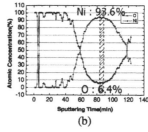
(b)

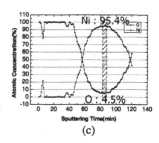
(c)

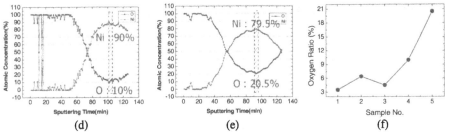

| (d) | (e) | (f) |

Fig. 1 AES analysis of sample (a) #1, (b) #2, (c) #3, (d) #4, (e) #5. (f) Oxygen ratio by samples.

These results show that the Ni to oxygen ratio in the middle layer was 96.5:3.5 in sample #1 and 93.6:6.4 in sample #2. Therefore, when O_2 plasma oxidation was not used, the oxygen content was below 6.5%. Samples #3, #4, and #5 had O_2 plasma oxidation, and the Ni to oxygen ratio was 95.4:4.5, 90:10, and 79.5:20.5, respectively. A comparison of samples #5 and #2, which had the same Ni evaporation rate but one with and one without O_2 plasma oxidation, showed the oxygen content of sample #5 was about 3.2 times that of sample #2. This conclusively showed that when the evaporation rate of the Ni layer was low, preferably less than 0.1 Å/sec, and O_2 plasma oxidation was used, the oxygen ratio in the Ni nanocrystal layer increased as shown in Fig. 1(f).

X-ray photoelectron spectroscopy (XPS) analysis

We analyzed the samples by XPS (Fig. 2) to confirm the results of the AES analysis.

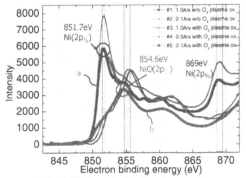

Fig. 2 XPS analysis of each sample

A comparison of the XPS results for samples #1 and #5 showed the biggest oxygen content difference to the AES analysis. In Fig. 2, sample #1 (a) displays a peak at 851.7 and at 869 eV, where the Ni nanocrystal layer was composed of Ni(2p$_{3/2}$) and of Ni(2p$_{1/2}$). An Ni peak shift to 854.6 eV corresponds to NiO(2p$_{3/2}$) in sample #5 (b). This is evidence that O_2 plasma oxidation effectively oxidizes Ni.

Transmission electron microscope (TEM) analysis

We did a TEM physical analysis in addition to the chemical composition analysis. x-TEM images of samples #1 and #5 are shown in Figs. 3(a) and (b), and high-resolution TEM images are shown in Figs. 3(c) and (d).

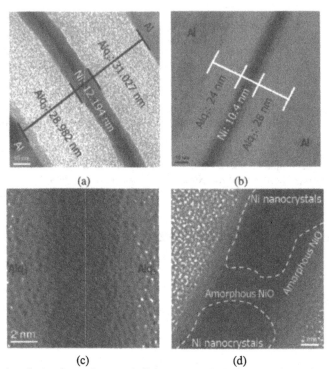

(a)   (b)

(c)   (d)

Fig. 3 Physical analysis of organic nonvolatile memory. x-TEM images of sample (a) #1, (b) #5. High-resolution TEM images of samples (c) #1, (d) #5

Figure 3(c) shows a crystal fringe around the whole Ni nanocrystal layer. The nanocrystals have a face-centered cubic crystal structure of pure poly Ni according to result that analyze by diffraction pattern analysis of TEM. However, in Fig. 3(d), the amorphous NiO, which has no crystal fringe, is between nanocrystals with crystal fringe patterns in various directions. This NiO was expected and could be inferred from the AES and XPS analyses. As a result, we confirmed that the formation conditions for sample #5 were the best for isolating the nanocrystals with amorphous NiO.

Electrical characteristics

Last, we measured the I-V characteristics (Fig. 4). As expected because of the previous analyses results, we obtained the best bistable switching characteristics in sample #5 (Fig. 4(b)).

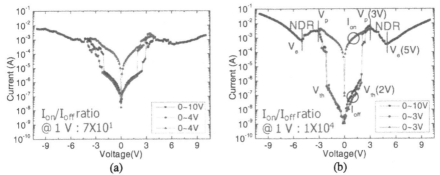

Fig. 4 Electrical characteristics of organic nonvolatile memory. I-V characteristics of sample (a) #1, (b) #5

Basically, the current abruptly increased with an applied bias from V_{th} (threshold voltage: 2 V) to V_p (program voltage: 3 V) and then decreased with an increasing applied bias from V_p to V_e (erase voltage: 5 V), showing a negative differential resistance (NDR) region. Finally, the current slightly increased with an applied bias above V_e. After the first applied voltage was swept from 0 to 10 V, the second applied bias was swept from 0 to V_p, where the current followed a high-resistance state (I_{off}). After the second applied bias was swept, the third applied bias (reading after program) was swept from 0 to V_p again, where the current followed a high-current state or low-resistance state (I_{on}). In a comparison of the results from sample #5 in Fig. 4(b) with those of sample #1 in Fig. 4(a), the V_{th}, V_p, and V_e of sample #5 decreased from 2.5, 4, and 6.3 V to 2, 3, and 5 V, respectively. Moreover, the I_{on}/I_{off} ratio (memory margin) increased by about 140 times, from 70 to 10^4.

CONCLUSION

This study shows the process to optimize the formation of the nanocrystal layer, which is important for revealing the bistable switching and memory characteristics of organic nonvolatile memory. We obtained well isolated nanocrystals with a high oxygen content by in-situ O_2 plasma oxidation after depositing the Ni at a low evaporation rate. Thereby, we developed good organic nonvolatile memory having a big I_{on}/I_{off} ratio (memory margin). Our chemical and physical analysis and process optimization is an effective method for obtaining wanted nonvolatile memory characteristics in the nanocrystal layer even when other metals are used.

ACKNOWLEDGEMENTS

This research was supported by the national research program for 0.1-Terabit Non-volatile Memory Development sponsored by the Korean Ministry of Commerce, Industry, and Energy.

REFERENCES

1. J.G. Park, G. S. Lee, K. S. Chae, Y. J. Kim, T. Miyata, J. Kor. Phys. Soc. **48**, 1505 (2006)
2. L. P. Ma, J. Liu, Y. Yang, Appl. Phys. Lett. **80**, 2997 (2002)
3. L. D. Bozano, B. W. Kean, V. R. Deline, J. R. Salem, J. C. Scott, Appl. Phys. Lett. **84**, 607 (2004)
4. L. P. Ma, S. M. Pyo, J. Ouyang, Q. Xu, Y. Yang, Appl. Phys. Lett. **82**, 1419 (2003)
5. L. D. Bozano, B. W. Kean, M. Beinhoff, K. R. Carter, P. M. Rice, J. C. Scott, Adv. Funct. Mater. **15**, 1933 (2005)
6. Y. Yang, J. Ouyang, L. P. Ma, R. J. Tseng, C. W. Chu, Adv. Funct. Mater. **16**, 1001 (2006)
7. B. O. Cho, T. Yasue, H. S. Yoon, M. S. Lee, I. S. Yeo, U. I. Chung, J. T. Moon, B. I. Ryu, Electron Devices Meeting 2006, IEDM (2006)
8. C. W. Chu, J. Ouyang, J. H. Tseng, Y. Yang, Adv. Mater. **17**, 1440 (2005)
9. D. Tondelier, K. Lmimouni, D. Vuillaume, Appl. Phys. Lett. **85**, 5763 (2004)
10. D. Ma, M. Aguiar, J. A. Freire, I. A. Hümmelgen, Adv. Mater. **12**, 1063 (2000)

Mater. Res. Soc. Symp. Proc. Vol. 1071 © 2008 Materials Research Society   1071-F09-18

Current Conduction Mechanism for Non-volatile Memory Fabricated with Conductive Polymer Embedded Au Nanocrystals

Jong Dae Lee, Hyun Min Seung, Byeong Il Han, Gon-Sub Lee, and Jea-gun Park
Electrical & Computer Engineering, Hanyang University, Nano SOI Process Laboratary, Room #101, HIT, 17 Haengdang-dong, Seoungdang-gu, Seoul, 133-791, Korea, Republic of

ABSTRACT

Molecular memory is an expected next-generation, nonvolatile memory because it demonstrates the characteristic of a bi-stable switch and has a 45-nm initial feature, an up to 10-ns access and store time, and low-cost, flexible, and simple fabrication. Several types of molecular devices have been reported, such as simple and low molecular organic devices and polymer devices. However, fabricating this device is complex because the $Ni_{1-x}F_x$ crystals are embedded in the polyimide layer as a floating gate in the flash memory after defining the source and drain regions. We report the memory effects, based on the electrical bistability of the materials, in organic molecules. A bistable phenomenon was observed in a poly(N-vinylcarbazole) (PVK) layer, which contained a high density of Au nanocrystals and was sandwiched between Al electrodes without source and drain regions. The memory phenomenon in this device was based on the electrical bistability of the material, which has two resistance states. Nonvolatile memory devices that use discrete nanocrystals as charge storage sites and exhibit bistability have also been reported. We discuss the current conduction mechanism for nonvolatile memory devices.

INTRODUCTION

Organic nonvolatile memory devices have been investigated as candidates for next-generation nonvolatile memory because of their low-cost, flexible, and simple fabrication [1-5]. The memory phenomenon in these devices is based on the electrical bistability of the material, which has two resistance states [6]. We report on this phenomenon in the organic molecules and polymer nonvolatile memory device. A bistable phenomenon was observed in the poly(N-vinylcarbazole) (PVK) layer, which contained a high density of Au nanocrystals and was sandwiched between Al electrodes. This device showed good nonvolatile memory characteristics. We suggest that the current conduction mechanism for nonvolatile memory fabricated with conductive polymer embedded Au nanocrystals clearly follows space-charge-limited current (SCLC) for a low conductivity state, thermionic field emission for electron charge (writing) or discharge (erasing), and F-N tunneling after erasing.

EXPERIMENT

A 600-nm HDPCVD film was deposited on an SC1-cleaned p-type silicon substrate. A solvent (chloroform) was used to dissolve the PVK. The weight percent was varied by altering the viscosity and thickness of the PVK film. An 80-nm bottom Al electrode was thermally evaporated at a 10^{-4}-Pa chamber pressure and a 5-Å/s evaporation rate using a shadow mask. The first PVK layer was deposited by spin coating with a 2000-rpm rotation velocity for 99 s and then baking at 120°C for 2 min to evaporate the solvent. An Au film was thermally evaporated at a 5×10^{-6}-Torr chamber pressure and a 0.1 to 0.3-Å/s evaporation rate to a 5-nm thickness using a shadow mask with $4F^2$, and a second PVK layer was deposited by spin coating with a 2000-rpm rotation velocity for 99 s and then baking at 120°C for 2 min. Next, it was cured at 300°C for 2 h to produce Au nanocrystals. Finally, the top Al electrode was thermally evaporated at a 10^{-4}-Pa chamber pressure and a 5-Å/s evaporation rate to an 80-nm thickness using a shadow mask that was vertically patterned against the bottom electrode (Fig. 1).

Figure 1. Device and chemical structure analysis of polymer PVK nonvolatile $4F^2$ memory cell fabricated by curing.

DISCUSSION

Our device, fabricated with conductive polymer embedded Au nanocrystals, showed good nonvolatile memory characteristics. Figure 2 shows its typical bistable characteristics, with several regions of current levels (I_{on}, I_{off}, I_{inter}). When the voltage was increased from zero in a low conductivity (I_{off}) state, the current increased rapidly at the threshold voltage (V_{th}) and presented a regime of negative differential resistance (NDR) after writing. Moreover, I_{on} and I_{off} states could be set at $V_{program}$ (or V_p) and V_{erase} (or V_e), respectively, and could be read at 2 V. After the device was programmed by sweeping the voltage from 0 to V_p, the current followed a high conductivity state and stayed in the I_{on} state. When the device was programmed by sweeping the voltage from 0 to V_e, the current followed a low conductivity state and stayed in the I_{off} state. Furthermore, by sweeping the voltage from 0 to V_{NDR}, the device exhibited seven different reversible current paths (intermediate (I_{inter}) states) enabling electron charges and discharges on the surface of the Au nanocrystals. Our results show that the fundamental parameters of the device were stable; the values of V_{th}, V_p, and V_e were respectively 2.8, 4, and 8 V, respectively. In particular, this device exhibited excellent nonvolatile memory behavior, with a bistability (I_{on}/I_{off}) of $>1 \times 10^2$ and an intermediate (I_{inter}) state for multi-bit operation [7].

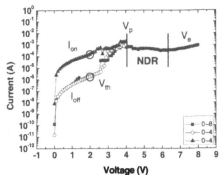

Figure 2. Electrical analysis of polymer PVK nonvolatile $4F^2$ memory cell fabricated by curing: DC I-V curve.

Mechanism

Most of the experimental features presented here are identical to the features of an SV model [8]. We hypothesized that the mechanism inducing the bistability consisted of charge trapping and of space-charge field inhibited electron transfers, so we used equations (Table 1) to fit the I-V curves to demonstrate the current conduction mechanism. When the voltage was swept from 0 to V_{th} (threshold voltage) for a low conductivity state, the mechanism followed SCLC. When region of NDR that V_{th} to V_p (program voltage) for programming and V_p to V_e (erase voltage), the mechanism followed thermionic field emission and combined F-N tunneling and thermionic field emission for an electron charge on the Au nanocrystals. When V_e was swept to 8 V for erasing, the mechanism followed F-N tunneling for an electron discharge on the nanocrystals. Finally, when the voltage was swept from 0 to V_{th} for a high conductivity state, the mechanism combined SCLC and thermionic emission.

Space-charge-limited current	$J_{SCLC} = \dfrac{8\varepsilon_i \mu V^2}{9d^3}$
F-N tunneling	$J_{FN} \approx V^2 \exp(-b/V)$
Thermionic emission	$J \propto T^2 \exp[\dfrac{q}{kT}(a\sqrt{V} - \phi_B)]$
Thermionic field emission	$J_{TFE} = \dfrac{AT}{k}\sqrt{\pi E_{00} q \left[V_R + \dfrac{\phi_{Bn}}{\cosh^2(E_{00}/kT)}\right]}$ $\cdot \exp\left(\dfrac{-q\phi_{Bn}}{E_0}\right)\exp\left(\dfrac{qV_R}{\varepsilon'}\right)$

Table 1. Equations for fitting I-V curves to confirm current conduction mechanism for nonvolatile memory fabricated with conductive polymer embedded Au nanocrystals.

155

Cross-sectional TEM analysis

We observed our 4F² memory cell physically and chemically. An x-TEM image (obtained at a 200-kV acceleration voltage) of one of the cells we fabricated is shown in Fig. 3(a) The thicknesses of the top PVK, middle Au, and bottom PVK layers were, respectively, about 32.482, 8 to 9.5, and 37.404 nm. A magnified x-TEM image (obtained at a 600-kV acceleration voltage) of the cell shown in Fig. 3(a) is shown in Fig. 3(b), where the diameter of the uniformly distributed Au nanocrystals is about 8 to 9 nm. The nanocrystals were well isolated from one another.

(a)   (b)

Figure 3. Physical analysis of polymer PVK nonvolatile 4F² memory cell: (a) x-TEM image (at 200-kV acceleration voltage), (b) magnified x-TEM image (at 600-kV acceleration voltage).

CONCLUSION

In summary, the nonvolatile memory fabricated with conductive polymer (PVK) embedded Au nanocrystals by a curing process showed good nonvolatile memory characteristics. To understand the mechanism observed in this device's I-V curve sweep, we used equations to fit the curve and discovered that the current conduction mechanism clearly followed space-charge-limited for a low conductivity state, thermionic field emission for electron charge (writing) and discharge (erasing), and F-N tunneling after erasing. This type of device is easy to fabricate and shows promise as a next-generation nonvolatile memory device.

ACKNOWLEDGMENTS

This research was supported by the national research program for 0.1-Terabit Non-volatile Memory Development sponsored by the Korean Ministry of Commerce, Industry, and Energy.

REFERENCES

1. L. D. Bozano, B. W. Kean, M. Beinhoff, K. R. Carter, P. M. Rice, and J. C. Scott, Adv. Funct. Mater. **15**, 1933 (2005).
2. C. W. Tang and S. A. Van Slyke, Appl. Phys. Lett. **51**, 913 (1987).

3. J. H. Burroughs, D. D. C. Bradley, A. R. Brown, R. N. Marks, K. Mackay, R. H. Friend, P. L. Bums, and A. B. Holmes, Nature (London) **347**, 539 (1990).
4. F. Garnier, R. Hajlaoui, A. Yassar, and P. Srivastava, Science **265**, 1684 (1994).
5. N. Tessler, G. J. Denton, and R. H. Friend, Nature (London) **382**, 695 (1996).
6. L. Ma, S. Pyo, J. Ouyang, Q. Xu, and Y. Yang, Appl. Phys. Lett. **82**, (2003) 1419.
7. L. D. Bozano, B. W. Kean, V. R. Deline, J. R. Salem, and J. C. Scott, Appl. Phys. Lett. **84**, 607 (2004).
8. J. G. Simmons and R. P. Verderber, Proc. R. Soc. London, Ser. A **301**, 77 (1967).

Organic Ferroelectric Memory

Mater. Res. Soc. Symp. Proc. Vol. 1071 © 2008 Materials Research Society 1071-F03-10

Organic Field Effect Transistor using BaTiO3-Mn Doped and P(VDF-TrFE) for Non Volatile Memory Application

Sambit Pattnaik, Ashish Garg, and Monica Katiyar
Department of Materials and Metallurgical Engineering, Indian Institute of Technology Kanpur, Kanpur, 208016, India

ABSTRACT

Here, we report fabrication of an organic field effect transistor that can be used as a memory device. We have evaluated inorganic ferroelectric insulator manganese doped barium titanate(BTO), organic poly(vinylidene fluoride trifluoroethylene) P(VDF-TrFE), and their composite. The inorganic and organic ferroelectrics were fabricated using low cost process of spin coating followed by annealing to enhance crystallinity. The ferroelectric phase evolution is assessed by X-ray diffraction, MIM structure is used to study polarization behaviour and leakage current. Finally, OFETs are fabricated using thermal evaporation of 75 nm of pentacene. Gold electrodes of 70 nm were evaporated for the top contact devices keeping W/L=40. The OFET devices, for BTO/P(VDF-TrFE) composite insulator, showed memory effect with shift in threshold voltage of 8.5 ± 1.5V.

INTRODUCTION

For Last two decades there has been a surge in the field of organic based devices in the area of organic solar cells, light emitting diodes and field effective transistors (OFETs). Recent advancements in organic field effect transistors (OFETs) have shown potential for making low cost memory devices. The basic model of a ferroelectric nonvolatile memory is to use a ferroelectric capacitor driven by one or more transistors, but this structure can be integrated by using a ferroelectric material as the gate insulator of field effect transistor. The remnant polarization of the ferroelectric material can be used to control the surface state of the semiconductor [1]. More recently there have been a lot of research efforts in the field of organic memories based on ferroelectric polymers, most of the work has been concentrated on P(VDF-TrFE) copolymer and ferroelectric OFET's have been demonstrated by various groups [2-5]. But the voltages required for these devices to operate are very high and practical application not feasible, so there is a need to lower the operating voltages.

In this paper, we report use of inorganic, organic, and composite dielectric using low cost process of fabrication. The composite is fabricated by using two ferroelectric materials; an inorganic BTO (BaTi$_{0.995}$Mn$_{0.05}$O$_3$) and organic P(VDF-TrFE) (70-30 mol %). First the conditions have been optimized for the dielectrics and then OFETs have been fabricated to see memory performance.

EXPERIMENTAL DETAILS

The inorganic dielectric has been made by spin coating of $BaTi_{0.995}Mn_{0.05}O_3$ (BTO) on clean Pt (111)/TiO_2/SiO_2/Si substrates. The substrates were cleaned by first ultra sonication in acetone followed by propanol and finally in deionised water and then drying at 200°C. The BTO films were deposited by chemical solution deposition from a solution of 0.4M BTO prepared by dissolving 98.5% assay barium acetate in glacial acetic acid, 97% Ti-isopropoxide in acetyl acetone and 99% manganous acetate in glacial acetic acid in stoichiometric quantities and stirring them together for 6 hrs with 2- methoxy ethanol to stabilize the solution to store after filtering through a 0.2μm filter. Multiple layers of BTO were deposited on clean Pt/Si substrates to achieve different thicknesses with each layer step pyrolysed to remove volatile organics and finally annealed to enhance crystallinity. To characterize the ferroelectric oxide gold electrodes of $0.5mm^2$ were deposited to form a MIM structure as in fig.1 (a).

P(VDF-TrFE) was dissolved in methyl ethyl ketone at different concentration to achieve various thicknesses and filtered through a 0.2μm filter. Indium tin oxide coated glass (ITO) was first cleaned by RCA process and then a smoothening layer of 70nm of PEDOT: PSS was spin coated on to the ITO after which P(VDF-TrFE) was spin coated and then dried at 90°C in air to evaporate the solvent. Later the film was annealed in vacuum to get a semi crystalline phase subsequently gold electrodes were deposited to form MIM structure for characterizing the organic ferroelectric.

The composite dielectric was formed by first spin coating BTO on Pt-Si followed by spin coating of P(VDF-TrFE) on the BTO film by the above mentioned process. Finally OFET structure was completed as shown in figure 1(b) by evaporation of pentacene in a vacuum of 2 x 10^{-6} mbar at 1.5-2.0 nm/min at a substrate temperature of 60°C. Top contacts electrodes were deposited by gold evaporation through shadow masking to form devices with a channel length of 25μm and W/L=40. The dielectrics were characterized for phase evolution using Grazing angle X-Ray Diffraction. Ferroelectric characterization was done using Radiant ferroelectric tester and capacitance measurements were carried using Agilent 4294A impedance analyzer and finally transistor characterization and memory effect was confirmed on the OFET structure using Keithley 4200s semiconductor parameter analyzer in air. The surface morphology and roughness was measured with NT-MDT Atomic Force Microscope.

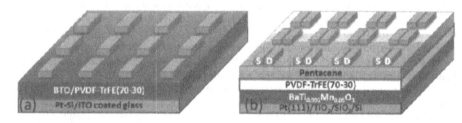

Figure 1: (a) MIM structure and (b) OFET structure.

RESULTS AND DISCUSSION

Fig. 2(a) shows X-ray diffraction pattern of the BTO films grown on Pt/Si, the pattern exhibits the phase evolution of polycrystalline $BaTiO_3$ formed by annealing at 800°C for 2mins.

Fig. 2(b) shows the formation of semi crystalline ITO/P(VDF-TrFE) by annealing in a vacuum oven at 135°C for 2 hours. The BTO film made by chemical solution deposition could not be used directly as gate insulator as it showed cracks as seen on optical micrograph shown in fig.2(c) where as P(VDF-TrFE) was relatively smooth with a rms roughness of 2.75±0.2nm as shown in fig. 2(d). Fig. 2(e) shows the morphology of the composite dielectric of BTO/P(VDF-TrFE).

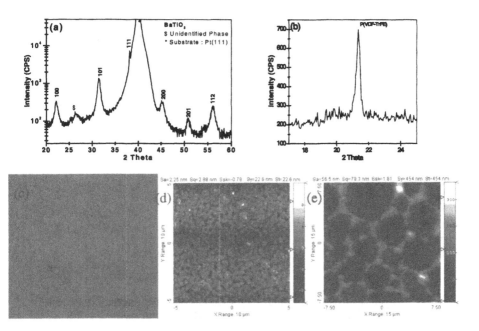

Figure 2: XRD at a grazing angle of 2° of (a) BTO (b) P(VDF-TrFE) (c) shows optical micrograph of BTO at 1000X and (d) morphology of P(VDF-TrFE) (e) morphology BTO/P(VDF-TrFE) composite dielectric

The ferroelectric characterization was done on the following MIM structures as shown in fig. 3(a). The figure shows high polarization of BTO but with leaky hysteresis with a remnant polarization of $10\mu C/cm^2$. In case of the organic ferroelectric polymer P(VDF-TrFE) the ferroelectric polarization showed good saturation and high remnant polarization of $8\mu C/cm^2$ shown in fig. 3(b). Here, we observed a large difference between the polarization states which may be due to slower movement of charges in organic materials or may be due to some trap charges that exist in the interfaces. Both ferroelectric characterizations were done at 1 KHz. The leakage current through the dielectric plays a major role in determining the working of an OFET as the current would like to take the easiest path present and if the leakage currents are high then most of the current are going to leak away from the gate rather than through the semiconductor. The leakage current measurements were done on the MIM structures as shown in figure 3(c) and

3(d) for Pt-Si/BTO/Au and ITO/PEDOT: PSS/P(VDF-TrFE)/Au, respectively. The leakage currents obtained for both cases were on the higher side and were too high for OFET fabrication.

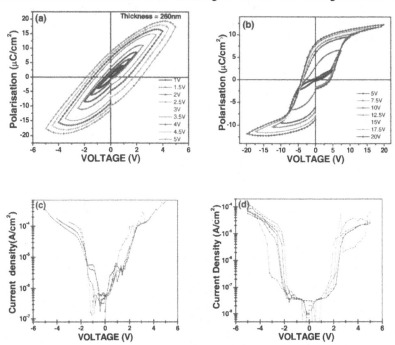

Figure 3: Ferroelectric hysteresis of (a) BTO and (b) P(VDF-TrFE) and leakage currents of various devices of (c) BTO and (d) P(VDF-TrFE) in MIM structure.

Therefore, a composite dielectric was fabricated which would fill up the cracks that have been formed on the BTO film by P(VDF-TrFE) and lead to decrease in the leakage currents. Also the semiconductor-dielectric plays a major role in OFETs as conduction takes place in first few mono layers near the interface so an organic-organic interface could be better. In fig 2(e) we can see that there is clear indication P(VDF-TrFE) segregation which may be at the grain boundaries of BTO and thus able to decrease the leakages taking place through the gate formed by composite dielectric of BTO/P(VDF-TrFE). So an OFET with Pt-Si/BTO/P(VDF-TrFE)/pentacene/Au was fabricated and analyzed. The capacitance of the Pt-Si/BTO/P(VDF-TrFE)/pentacene/Au varied between 25 nF/cm^2 to 10 nF/cm^2 with increasing frequency. The output characteristics of the OFET structures have been shown in fig. 4(a) and the memory effect in the OFET is confirmed by the transfer characteristics as shown in fig. 4(b) which shows that there is a change in the threshold voltage of around 8± 1V with an I_{on}/I_{off} ratio of 10^2. This value is rather low possibly due to the low mobilities of the order of ~4±1 x10^{-3} cm^2/V.s which can be related to the high roughness obtained on BTO/P(VDF-TrFE) of around 75±10nm shown in fig. 2(e) which resulted in a bad interface quality between the dielectric-semiconductor as well as the surface roughness of deposited pentacene was at much higher order of 45±5nm.

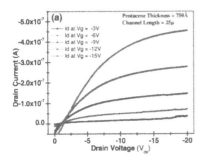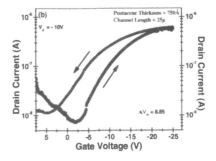

Figure 4: (a) Output characteristics OFET and (b) Transfer characteristics of OFET using composite insulator BTO/P(VDF-TrFE).

CONCLUSIONS

In summary we have been able to fabricate a memory device in the form of an OFET structure. These device structures resulted in decreasing the operating voltages for the memory window as well as decrease the gate leakages but their working is limited due to the low range of mobility which is mainly dependent on the interface between the dielectric-semiconductor. There is scope for further improvement of the mobility by surface treatment and optimization of deposition parameters of pentacene which would lead to higher I_{on}/I_{off} ratios and hence a much better memory window.

ACKNOWLEDGEMENT

The authors would like to thank Samtel center for display technologies for their facilities and Solvay-Solexis for providing us with P(VDF-TrFE).

REFERENCES

1. Hand book of thin films Devices, edited by Maurice. H. Francombe, Vol. 5: Ferroelectric Thin Films, 79-89 (2000).
2. R. C. G. Naber, J. Massolt, M. Spijkman, K. Asadi, P. W. M. Blom and D. M. de Leeuw, Applied Physics Letters **90** (11) (2007).
3. R. C. G. Naber, M. Mulder, B. de Boer, P. W. M. Blom and D. M. de Leeuw, Organic Electronics **7** (3), 132-136 (2006).
4. R. C. G. Naber, C. Tanase, P. W. M. Blom, G. H. Gelinck, A. W. Marsman, F. J. Touwslager, S. Setayesh and D. M. De Leeuw, Nature Materials **4** (3), 243-248 (2005).
5. C. A. Nguyen, P. S. Lee and S. G. Mhaisalkar, Organic Electronics **8** (4), 415-422 (2007).

Mater. Res. Soc. Symp. Proc. Vol. 1071 © 2008 Materials Research Society   1071-F03-05

Fabrication and Characterization of Ferroelectric Polymer/TiO$_2$/Al-doped ZnO Structures

Koji Aizawa
OEDS R&D Center, Kanazawa Institute of Technology, 7-1 Ohgigaoka, Nonoichi, Ishikawa, 921-8501, Japan

ABSTRACT

Fabrication and characterization of 700-nm-thick poly(vinylidene fluoride/trifluoroethylene) [P(VDF/TrFE)]/TiO$_2$/Al-doped ZnO (AZO) structures on a glass substrate were introduced. In this study, the sputtered TiO$_2$ films as insulator were used for the reduction of leakage current. The leakage current density of the fabricated Pt/P(VDF/TrFE)/170-nm-thick TiO$_2$/AZO and Pt/P(VDF/TrFE)/AZO structures were approximately 3.9 and 8.7 nA/cm^2 at the applied voltage of 10 V, respectively. In the polarization vs. voltage characteristics, the fabricated Pt/P(VDF/TrFE)/TiO$_2$/AZO structures showed hysteresis loops caused by ferroelectric polarization. The remnant polarization 2P$_r$ and coercive voltage 2V$_C$ measured from a saturated hysteresis loop at the frequency of 50 Hz were approximately 12 μC/cm^2 and 105 V, respectively. These results suggest that the insertion of TiO$_2$ film is available for reducing the leakage current without changing the ferroelectric properties of the P(VDF/TrFE) film.

INTRODUCTION

Ferroelectric-gate field effect transistors (FeFETs) have attracted much attention for memory applications because of nonvolatile data retention and nondestructive data readout [1]. Up to now, ferroelectric oxides such as Pb(Zr, Ti)O$_3$ and SrBi$_2$Ta$_2$O$_9$ were used as an insulator of FeFET [2,3]. However, these films were deposited at the temperature above 500 °C in order to grow the ferroelectric phase and realize the good ferroelectric properties. On the other hand, the ferroelectric polymer as a gate insulator of the FeFET was one of the promising candidates for nonvolatile memory applications [4]. It is expected that the ferroelectric polymer-gate FETs are widely used in the applications such as flat panel displays and flexible integrated circuits because the polymer films are deposited at low temperature on the plastic or glass substrates. Poly(vinylidene fluoride/trifluoroethylene) [P(VDF/TrFE)] copolymer is promising as ferroelectric polymer of the FeFETs due to large remnant polarization of 10 μC/cm^2 and low temperature growth under 140 °C [5,6]. In P(VDF/TrFE), TrFE is introduced in order to enhance the growth of the ferroelectric phase (β phase), while the leakage current of this polymer is increased by the leakage path formed in a TrFE molecule [6]. It is known that the gate leakage current deteriorates the device performance of the FeFETs. Recently, polymer dielectric/ferroelectric double-layer gate insulator was used for reducing the gate leakage in the FeFETs with ZnO film as a semiconductor [7]. In this study, characterization of the P(VDF/TrFE) films deposited on the TiO$_2$/Al-doped ZnO (AZO) structures was investigated. In these structures, the TiO$_2$ film is used as an insulator for reducing the leakage current.

EXPERIMENT

270-nm-thick AZO films were prepared on the glass substrates at the temperature of 200

°C by radio frequency (RF) magnetron sputtering apparatus (ULVAC) using a sintered AZO target (Tosoh Co., Ltd.). The AZO target was composed of ZnO and Al_2O_3 with a mass concentration of 2 wt%. The argon (Ar) gas pressure and RF power during deposition of the AZO films were 0.2 Pa and 100 W, respectively. The resistivity and hall mobility of the deposited AZO films were approximately 10^{-3} Ωcm and 20 cm^2/Vs. TiO_2 films were partially deposited on the AZO/glass substrates at room temperature through a stencil mask by RF magnetron sputtering using pure Ti target. Ti target was sputtered in a gas mixture of Ar and oxygen (O_2). The gas pressure and RF power during deposition of the TiO_2 films were 0.13 Pa and 200 W, respectively. The polymer films were prepared by employing a solution method with N,N-dimethylformamide (DMF) as solvent. P(VDF/TrFE) copolymer powder (Kureha) with composition of 75/25 mol% were used in this experiment. Its polymer was dissolved in DMF with a mass concentration of 10 wt%. A dissolved P(VDF/TrFE) solution was spin-coated on the AZO/glass substrates with and without TiO_2 film. The coated films were dried at 140 °C for 1 h in vacuum due to growth of ferroelectric phase. The film thickness of P(VDF/TrFE) film was approximately 700 nm. The crystallinity of the fabricated samples was analyzed by X-ray diffraction (XRD) using Cu Kα radiation (Rigaku). Finally, the Pt electrodes were formed through a stencil mask by dc sputtering at room temperature, in which the electrode area was approximately 3x10^{-3} cm^2. Typical fabricated Pt/P(VDF/TrFE)/TiO_2/AZO structure is shown in Fig. 1. Top and bottom electrodes were fabricated on the P(VDF/TrFE) and AZO film, respectively. The electrical properties were measured between a top and bottom Pt electrode. The ferroelectric properties were measured by a standard Sawyer-Tower circuit. The current density-voltage (J-V) characteristics were also investigated at room temperature using a semiconductor parameter analyzer (Agilent).

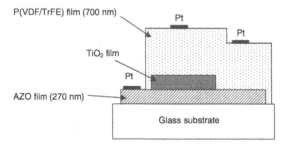

Figure 1 Illustration of Pt/P(VDF/TrFE)/TiO_2/AZO and Pt/P(VDF/TrFE)/AZO structures on a glass substrate. The thicknesses of the P(VDF/TrFE) and AZO films are approximately 700 nm and 270 nm, respectively. TiO_2 films are partially deposited on an AZO film through a stencil mask.

DISCUSSIONS
Leakage current of TiO_2/AZO structures

Figure 2(a) shows the J-V characteristics of the fabricated Pt/TiO_2/AZO structures with various TiO_2 thicknesses. The asymmetric J-V curves were observed because the work function of Pt (about 5.7 eV) is different from that of AZO (about 3.2 eV) [8,9]. The leakage current density and the resistivity of Pt/170-nm-thick TiO_2/AZO structures were approximately 9.6×10^{-7} A/cm^2 and 3.0×10^{11} Ωcm, respectively, when a negative bias voltage of -5 V was applied to the Pt electrode. Figure 2(b) shows the film thickness dependence of the leakage current density when an applied bias voltage is -5 V. Ar and O_2 gases during deposition of the TiO_2 films were introduced at the gas flow rate of 5 and 2 sccm, respectively. It can be seen from this figure that the leakage current density is reduced by increasing the film thickness. The measured current density J can be fitted by $J \sim t^{-4.5}$, where t is film thickness of TiO_2. In general, the space-charge-limited current in a trap-free insulator is described by $J \sim t^{-3}$ [10]. Therefore, this result shows that the leakage current of the Pt/ TiO_2/AZO structure has strong dependence of film thickness.

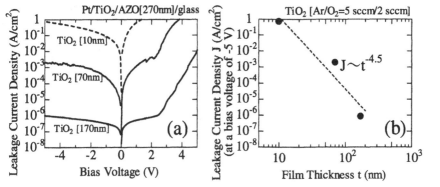

Figure 2 (a) J-V characteristics and (b) the thickness dependence of the leakage current of Pt/TiO_2/AZO structures with various TiO_2 thickness. A dashed line indicates the fitting of the relation between current density and film thickness.

Leakage current of P(VDF/TrFE)/TiO_2/AZO structures

Figure 3 shows the J-V characteristics of the fabricated Pt/P(VDF/TrFE)/TiO_2/AZO and Pt/P(VDF/TrFE)/AZO structures, in which the film thickness of TiO_2 is 170 nm. The leakage current density of the fabricated samples with and without TiO_2 films were approximately 3.9 and 8.7 nA/cm^2 at the applied voltage of 10 V, respectively. To investigate these results in detail, the electrical resistance of the P(VDF/TrFE) and TiO_2 films were estimated. The applied voltage V_{FE} of P(VDF/TrFE) and V_i of TiO_2 film are given by the following equations when the spontaneous polarization of the P(VDF/TrFE) is negligible.

$$V_{FE} = \frac{t_{FE} k_i}{t_{FE} k_i + t_i k_{FE}} \cdot V \tag{1}$$

$$V_i = \frac{t_i k_{FE}}{t_{FE} k_i + t_i k_{FE}} \cdot V \tag{2}$$

$$V = V_{FE} + V_i \tag{3}$$

Where k_{FE} (about 11) and t_{FE} and k_i (about 60) and t_i are the low frequency dielectric constants and the film thicknesses of P(VDF/TrFE) and TiO$_2$, respectively [11,12].

From these equations, the calculated values of V_{FE} and V_i are approximately 9.6 and 0.4 V, respectively, when an applied voltage V is 10 V. The calculated resistance of the 700-nm-thick P(VDF/TrFE) film is approximately 4×10^{11} Ω, when the J-V characteristics of the Pt/P(VDF/TrFE)/AZO structure are used in this estimation. On the other hand, the calculated resistance value of 170-nm-thick TiO$_2$ is approximately 2×10^9 Ω when the resistivity (3×10^{11} Ωcm) estimated from J-E characteristics of the Pt/TiO$_2$/AZO structures is used. Consequently, it seems that effect of the insertion of the TiO$_2$ film on the reduction of the leakage current is weak, because the electrical resistance of P(VDF/TrFE) film is about two times larger than that of TiO$_2$ film. However, it is clarify from Fig.3 that the insertion of the TiO$_2$ films is available for reducing the leakage current. As another view point, the interface effect should be considered to understand the current flow in the layered-structure. It is known that the conduction of the P(VDF/TrFE) film at a low electric field is mainly determined by Schottky emission at the P(VDF/TrFE)/metal interface [13]. Since the current flows through the P(VDF/TrFE)/TiO$_2$ interface, it is supposed that the leakage current of the Pt/P(VDF/TrFE)/TiO$_2$/AZO structures is also controlled by the conduction at the P(VDF/TrFE)/TiO$_2$ interface as well as those at the Pt/P(VDF/TrFE) and TiO$_2$/AZO interfaces. Therefore, the further investigation about the conduction mechanism of the P(VDF/TrFE)/TiO$_2$ interface is required.

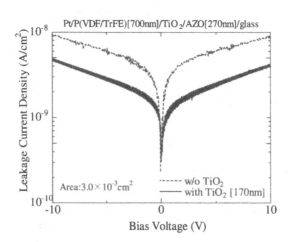

Figure 3 J-V characteristics of Pt/P(VDF/TrFE)/AZO and Pt/P(VDF/TrFE)/TiO$_2$/AZO structures. The film thicknesses of P(VDF/TrFE), TiO$_2$ and AZO are 700, 170 and 270 nm, respectively.

Ferroelectric property of P(VDF/TrFE)/TiO₂/AZO structures

From XRD analysis of the fabricated Pt/P(VDF/TrFE)/TiO₂/AZO structure, the diffraction peaks from the reflection of the crystalline β phase (ferroelectric phase) and c-axis-oriented AZO crystallites were observed. On the other hand, no diffraction peaks from TiO_2 crystallites were observed. Therefore, it is considered that the deposited P(VDF/TrFE) films exhibit the ferroelectric properties. Figure 4 shows the polarization vs. voltage (P-V) characteristics of the fabricated Pt/P(VDF/TrFE)/AZO and Pt/P(VDF/TrFE)/TiO₂/AZO structures, in which 50 Hz sinusoidal voltage signal was applied between a Pt and AZO film. It can be seen from P-V characteristics that the excellent hysteresis loops were observed in both samples. The remnant polarization $2P_r$ and coercive voltage $2V_C$ of the Pt/P(VDF/TrFE)/TiO₂/AZO structure were approximately 12 $\mu C/cm^2$ and 105 V (1.5 MV/cm as coercive force $2E_C$), respectively. The obtained $2P_r$ value was as same as that of the spin-coated P(VDF/TrFE) films reported in other literature [6]. On the other hand, the $2P_r$ value of the Pt/P(VDF/TrFE)/AZO structure was slightly larger than that of the Pt/P(VDF/TrFE)/TiO₂/AZO structure, because the applied voltage of P(VDF/TrFE) film was reduced by insertion of TiO_2 film. The calculated peak voltage applied to the P(VDF/TrFE) film in the Pt/P(VDF/TrFE)/TiO₂/AZO structure is approximately 80 V when the peak voltage of the measurement signal is 107 V. Consequently, the ferroelectric properties of the P(VDF/TrFE) film are not changed by introducing the TiO_2 film.

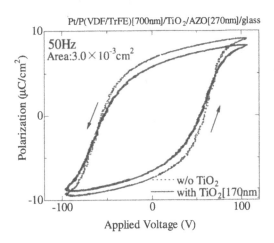

Figure 4 Typical 50 Hz P-V characteristics of the fabricated Pt/P(VDF/TrFE)/AZO and Pt/P(VDF/TrFE)/170-nm-thick TiO₂/AZO structures. Arrows show the hysteresis direction.

CONCLUSIONS

This work showed the electrical properties of 700-nm-thick P(VDF/TrFE)/TiO₂/AZO structures on a glass substrate and it was discussed that the effect of the TiO₂ insertion on the leakage current and ferroelectric properties. The leakage current density of the fabricated

Pt/P(VDF/TrFE)/170-nm-thick TiO_2/AZO structure was smaller than that of the Pt/P(VDF/TrFE)/AZO structure. In addition, the fabricated samples with and without TiO_2 film showed similar hysteresis loop in the P-V characteristics. In conclusion, the insertion of TiO_2 film is available for reducing the leakage current without changing the ferroelectric properties of the P(VDF/TrFE) film.

ACKNOWLEDGMENTS

The author especially thanks Y. Ohtani and K. Miyashita for their technical assistant in the experiments. This work was partly supported by a High-Tech Research Center Project Grant and Grant-in-Aid for Scientific Research (No. 18560316) from the Ministry of Education, Culture, Sports, Science and Technology, and by a Grant (Research for Promoting Technological Seeds, No. 06-014) from the Japan Science and Technology Agency.

REFERENCES

1. I. M. Ross, U.S.Patent No.2791760 (1957).
2. Y. Nakao, T. Nakamura, A. Kamisawa, and H. Takasu, Integr. Ferroelectr. 6, 23 (1995).
3. E. Tokumitsu, G. Fujii, and H. Ishiwara, Mat. Res. Soc. Symp. Proc. 493, 459 (1998).
4. R. C. G. Naber, C. Tanase, P. W. M. Blom, G. H. Gelinck, A. W. Marsman, F. J. Touwslager, S. Setayesh, and D. M. de Leeuw, Nat. Mater. 4, 243 (2004).
5. K. Tashiro, Ferroelectric Polymers, edited by H. Nalwa (Marcel Dekker, New York, 1995), p. 63.
6. S. Fujisaki, H. Ishiwara, and Y. Fujisaki, Appl. Phys. Lett. 90, 162902 (2007).
7. S. H. Noh, W. Choi, M. S. Oh, D. K. Hwang, K. Lee, S. Im, S. Jang, and E. Kim, Appl. Phys. Lett. 90, 253504 (2007).
8. H. B. Michaeson, J. Appl. Phys. 48, 4929 (1977).
9. T. Minami, MRS Bulletin 25, 38 (2000).
10. A. Rose, Phys. Rev. 97, 1538 (1955).
11. K. Muller, I. Paloumpa, K. Henkel, and D. Schmeisser, J. Appl. Phys. 98, 056104 (2005).
12. M. D. Stamate, Thin Solid Films 372, 246 (2000).
13. F. Xia and Q. M. Zhang, Appl. Phys. Lett. 85, 1719 (2004).

Mater. Res. Soc. Symp. Proc. Vol. 1071 © 2008 Materials Research Society 1071-F03-08

Fabrication and Electrical Studies of P(VDF/TrFE)(72/28) Copolymer based Non-Volatile Memory Devices as a Function of Varying Device Structures

Kap Jin Kim, Chang Woo Choi, Arun Anand Prabu, and Sun Yoon
Advanced Polymer and Fiber Materials, Kyung Hee University, 1 Seocheon-dong, Giheung-gu, Yongin, Gyeonggi-do, 446-701, Korea, Republic of

ABSTRACT

Ferroelectric characteristics of poly(vinylidene fluoride/trifluoroethylene) (P(VDF/TrFE) (72/28 mol%)) copolymer ultrathin films used as an insulator in varying memory device architectures such as metal-ferroelectric polymer-metal (MFM), MF-insulator-semiconductor (MFIS), MIS and organic field-effect transistor (OFET) were studied using different electrical measurements. A maximum data writing speed of 1.69 MHz was calculated from the switching time measured using MFM architecture. Capacitance-voltage measured using MFIS was found to be more suitable for distinguishing the '0' and '1' state compared to MFM device structure. In OFET, the decreasing channel length increased the measured drain current (I_d) values as well as its memory window enabling easier identification of the '0' and '1' state comparable to MFIS case. The data obtained from this study will be useful in the fabrication of non-volatile random access memory (NVRAM) devices with faster data R/W/E speed and memory retention capacity.

INTRODUCTION

The discovery of enhanced piezoelectric activity in poly(vinylidene fluoride) (PVDF) by H. Kawai in 1969 [1] and subsequent efforts by other researchers in the last few decades has led to significant progresses in PVDF and its copolymers for use in non-volatile random access memory (NVRAM) applications [2-5]. In particular, the copolymers of VDF and trifluoroethylene (TrFE) with VDF mole ratio between 50 to 80 % were reported by many researchers as a suitable ferroelectric insulator for use in NVRAM devices [4-8]. Among them, the P(VDF/TrFE) copolymer with VDF content 72 mol % exhibits reversible F⇔P phase transition behavior at T_c range, and hence has been intensely studied by many researchers including our group [6-11].

In our earlier studies, we have reported the changes in crystallinity, T_c behavior and C-F dipole orientation for P(VDF/TrFE)(72/28) samples as a function of varying thermal annealing [8] and film thickness [11] conditions. The data obtained from our previous studies were helpful towards understanding the ferroelectric characteristics of P(VDF/TrFE)(72/28) and for further realization of a workable memory device, the present work will focus more towards the lab-scale fabrication of a single-bit memory device based on P(VDF/TrFE)(72/28) ultrathin films. Our objective in the present work is to study the dipole switching behavior of P(VDF/TrFE)(72/28) copolymer nanoscale films as a function of varying device structures such as metal-ferroelectric polymer-metal (MFM), metal-ferroelectric polymer-insulating layer-semiconductor (MFIS) and organic field-effect transistor (OFET) using different electrical measurements. Our other objective is to reduce the writing voltage required for switching the varying devices used in this study by enhancing the ferroelectric properties of the P(VDF/TrFE)(72/28) insulator and the results are reported in detail here.

EXPERIMENTAL DETAILS

P(VDF/TrFE)(72/28 mol%) samples in the form of pellets were obtained from Solvay (U.S.A) and herein mentioned as VDF 72% copolymer. Methyl ethyl ketone (MEK), hexanol, polyvinyl alcohol (PVA), polystyrene (PS), poly-4-vinylphenol (PVP), and poly(melamine-co-formaldehyde) (PMF, used as PVP cross-linker) were purchased from Sigma-Aldrich (Korea). Au coated glass slides and SiO_2-Si wafer (used as bottom electrode) were purchased from SD Tech. (Korea) and Inostek (Korea), respectively.

VDF 72% copolymer of varying thickness were prepared by spin-casting different wt.-% of the samples in MEK solvent using a spin-coater (1500 rpm/30 s) under N_2 atmosphere over varying substrates like ITO on slide glass (for MFM and OFET), gold on slide glass (for MFM) and p-type Si-wafer (for MIS, MFIS and OFET). The thickness of the insulating layer in MFIS (PVP/PMF) was controlled by varying the solution concentration and setting the spin-coating speed at 3000 rpm/30 s. All the spun-cast samples were annealed (AN) above T_c (120°C, 3h in vacuum) to improve their degree of crystallinity and thermodynamic stability. Surface profilometer (Surfcorder ET-3000, Kosaka., Japan) was used to confirm the spin-coated sample thickness. For electrical studies, aluminum (Al) or gold (Au) as top electrode was deposited through the shadow mask under pressure of $\sim 10^{-6}$ torr on the spin-coated sample using a thermal evaporation system.

The polarization-electric field (P-E) hysteresis task is the most general and common tool for characterizing ferroelectric materials and its hysteresis was obtained by using a ferroelectric tester (Precision LC, Radiant Tech, U.S.A). In capacitance-voltage (C-V) measurement, frequency is applied with DC bias voltage between the gate electrode and substrate. External step bias generated by Keithley2400 was applied to LCR meter (ZM-2353) connected to a probe station similar to that used in P-E measurement. All these measurement process was controlled by LabVIEW program integrated with GPIB communication system. Transistor characteristics were measured at room temperature under dark conditions using a Hewlett-Packard 4156A semiconductor parameter analyzer equipped with an Agilent 16440A SMU/pulse generator.

DISCUSSION

P-E hysteresis studies

Among the device architectures used in this study, MFM and MFIS are simplest and the nonvolatile data writing capability of P(VDF/TrFE) is performed by electrically switching the C-F dipole under external electric field about two times larger than coercive field (E_c) (when $E>2E_c$, '1' state and when $E<-2E_c$, '0' state). In our earlier reports, the annealed VDF 72% copolymer samples were predicted to exhibit preferential molecular chain orientation (c-axis) along the substrate surface [8-10]. Thus, under an applied external electric field, the polarization direction of C-F dipoles (b-axis) can be changed selectively towards the electric field direction through cooperative dipole rotation along the chain, thereby leading to its large and permanent polarization.

With increasing electric field in the case of MFM structure (Figure 1(a) and (b)), remnant polarization (P_r) value also increases but reaches a saturation value beyond which even further increase in electric field does not induce any significant changes. Most of the P-E curves in MFM show asymmetric hysteresis which may be caused by the difference in chemical composition between top (aluminum) and bottom (ITO) electrodes, and hence the P_r and E_c are obtained from the average of $\pm P_r$ and $\pm E_c$, respectively. Compared to 100 nm (Figure 1(b)), the 50 nm sample as shown in Figure 1(a) exhibited much lower P_r value for the same bias voltage due to reduced degree of crystallinity. The chances of device shortage and breakdown also increased with reducing film thickness presumably due to increasing pinholes in 50 nm films as confirmed from our earlier study using AFM images [10]. P_r in thin films is mainly dependent on the crystallinity and the degree of chain and dipole orientation, which in turn is highly dependent on film preparation methods. Therefore it is very important to prepare VDF 72% copolymer ultrathin films with uniform film thickness and the highest possible P_r for use in NVRAM devices.

Using MFM device architecture, it is difficult to distinguish the '0 and 1' state without destroying the written data bit. As an example, for 100 nm VDF 72% sample (Figure 1(b)) with data bit written in '1' state ($+P_r$) and represented as 'B', the written data bit can be read by applying a gate bias voltage above its $+E_c$. However, for sample with data bit erased in the '0' state ($-P_r$) and marked as 'D', in order to read the '0' data bit, a gate bias voltage must be applied in the direction of A→B→C→D which will destroy the previous data state. In MFIS architecture case (Figure 1(c)), its P-E hysteresis loop does not present bistable behavior unlike MFM structure (Figure 1(b)). The P_r values were also diminished due to the presence of insulating layer (SiO_2) and Si-wafer in MFIS, though the advantage of using SiO_2 insulator is to avoid electrical shortage and to elevate thermal stability and retention time properties as well.

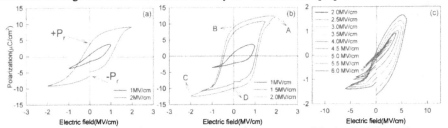

Figure 1. Polarization-electric field (P-E) hysteresis loops measured for VDF 72% copolymer (a) 50 nm and (b) 100 nm (MFM), and (c) 100 nm (MFIS).

From the above data, it is quite clear that the VDF 72% copolymer inherently exhibits both dielectric and ferroelectric characteristics which require varying time for switching the C-F dipoles when external voltage is applied. In the case of 200 nm VDF 72% sample, its switching time was measured to be 0.59 μs under 30V of external applied voltage. The switching time calculated for the above sample enabled us to expect a maximum data writing speed of 1.69 MHz and hence provide sufficient evidence towards the possibility of using VDF 72% copolymers for developing NvRAM devices. Also, with decreasing film thickness, the switching time was observed to be faster and the voltage required for switching the dipole also reduced significantly.

C-V hysteresis studies

Capacitance-applied voltage (C-V) pattern for VDF 72% samples using MFM and MFIS architecture are shown in Figure 2 (a) and (b), respectively. C-V curve in MFM case showed typical symmetrical loops having similar patterns to both positive and negative gate bias direction (Figure 2(a)). Previous researchers have reported the evaluation of MFIS and MIS based memory devices from the ratio of capacitance (C_{on}/C_{off}) when applied with cyclic bias voltage. From MFM symmetrical loops, it is not possible to identify the data bit state using the 'ON/OFF' ratio (C_{on}/C_{off}) at the zero gate bias. On the other hand, MFIS device using SiO_2 as an insulating layer was observed to exhibit asymmetrical loops, though shifting towards negative side as shown in Figure 2(b). The data bit state ('ON' and 'OFF') can be identified clearly with C_{on}/C_{off} ratio at -8V gate bias. However, in order to continuously maintain the written data state, it is necessary to apply -8V to the MFIS device and hence not a viable option in our attempt to find a workable NVRAM device structure.

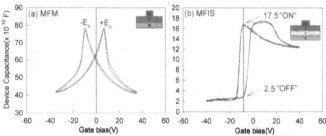

Figure 2. Capacitance-voltage (C-V) curve obtained from (a) Au/VDF 72% copolymer (100 nm) /Au (MFM) and (b) Au/VDF 72% copolymer (100 nm)/SiO_2 (100 nm)/n-Si (MFIS) device structures.

Fujisaki et. al. [12] suggested the fabrication of MFIS structure using Ta_2O_5 as an insulating layer to prevent the C-V hysteresis curve shifting towards negative gate bias. In our study, PVP used instead of SiO_2 as an insulating layer was found to prevent the shifting of C-V hysteresis curve in both MFIS and MIS towards the negative or positive gate bias voltage as shown in Figure 3(a) and (b), respectively. The maximum memory window width observed was ±40V and the width was considerably extended when using PVP as an insulator (Figure 3 (a)) compared to using SiO_2 (Figure 2(b)). Hence it is much easier to distinguish the 'ON' and 'OFF' data bit state at zero bias gate voltage, but writing and erasing voltages are increased so much. The memory window was observed to be smaller in the case of MIS structure and shifted to the positive voltage without using the ferroelectric polymer. However, the advantage of using MIS is that the operation voltage is much lower than that used in MFIS case.

To investigate the contribution of PVP in MFIS, we fabricated MFIS device with two different insulators, PVA (having hydroxyl group) and PS (having benzene group) as shown in Figure 3(c) and (d) respectively. Both -OH group in PVA and -C_6H_5 group in PS were observed to contribute towards preventing the shifting of C-V curve towards the negative side. However, the memory window widths observed in both the cases were found to be smaller than that observed with using PVP (Figure 3(a) and (b)) and hence further analysis is require to understand the mechanism.

Ferroelectric characteristics of OFET

Though there has been tremendous progress in the research of OFETs over the last two decades, very few attempts have been made to link the non-volatile memory element and the

OFET architecture. With decreasing channel length, the measured drain current (I_d) values increased as observed from Figure 4(a) and (d). From I_d-V_g hysteresis studies as shown in Figure 4(b), (c), (e) and (f), after writing with an applied gate voltage above positive coercive field (+E_c) of the ferroelectric polymer, the transistor works in the 'ON' state and the drain current at zero gate bias is considerably enhanced. In turn, after writing with an applied gate voltage above -E_c, the transistor goes to the 'OFF' state as expected. The drain current for the 'ON' state increased with decreasing channel length from 100 to 20 μm as shown in Figure 4(b) and (e) enabling easier identification of the '0' and '1' state comparable to the MFIS case. Similar trend was also noticed in the case of Figure 4(c) and (f), albeit in this case, the source-drain (S-D) width is higher than for samples shown in Figure 4(b) and (e). The increasing channel width also made a contribution towards increasing the ON/OFF current ratio of the OFET device.

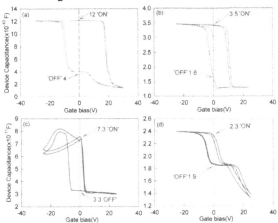

Figure 3. Capacitance-voltage (C-V) curve obtained from (a) Au/VDF 72% copolymer (100 nm) /PVP (130 nm)/p-Si (MFIS), (b) Au/PVP (130 nm)/p-Si (MIS), (c) Au/VDF 72% copolymer (200 nm)/PVA (80 nm)/p-Si (MFIS), and (d) Au/VDF 72% copolymer (200 nm)/PS (350 nm)/p-Si (MFIS) device structures.

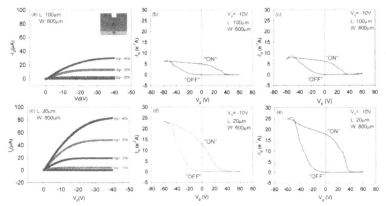

Figure 4. OFET using S-D (Au)/pentacene (40 nm)-PVP (80 nm)/VDF 72% copolymer (200 nm)/Gate (Au): I_d-V_d curves measured with varying S-D channel length (a) 100 μm and (d) 20 μm. OFET using S-D (Au)/pentacene (30 nm)/VDF 72% copolymer (180 nm)/PVP (80 nm)/Gate (p^{++} type Si): I_d-V_g hysteresis measured with varying S-D channel length: (b) 100 μm and (e) 20 μm (channel width=600 μm), and (c) 100 μm and (f) 20 μm (channel width=800 μm).

CONCLUSIONS

Electrical studies were useful in evaluating the ferroelectric and dipole switching behavior quantitatively for P(VDF/TrFE)(72/28 mol %) copolymer nanoscale films as a function of varying device architectures like MFM, MFIS, MIS and OFET. Though MFM is a simple device structure that can be effectively used to confirm the ferroelectric properties, it is not quite suitable for fabricating commercially viable memory devices due to the inconvenience faced during the data reading process as observed from P-E studies. From C-V measurements, MFIS and MIS devices were found to be more suitable for the easier identification of '0' and '1' state than MFM case. The disadvantage of using SiO_2 as an insulating layer was eliminated with using PVP layer which prevented the shifting of C-V hysteresis curve in both MFIS and MIS towards the negative gate bias voltage. In OFET, the contribution of increasing channel width and decreasing channel length towards increasing the memory window were noticed. The data obtained from this study were quite helpful in analyzing the suitability of using the material for NVRAM device applications.

ACKNOWLEDGMENTS

This work was supported by the SRC/ERC Program of KOSEF (R11-2005-065), the 0.1 Terabit Non-volatile Memory Development funded by the Ministry of Commerce, Industry, Energy of the Korean Government, and KRF international Academic Exchange Program (KRF-2006-D00021).

REFERENCES

1. H. Kawai, Jpn. J. Appl. Phys. **8**, 975 (1969).
2. J. T. Evans, and R. Womack, IEEE J. Solid-State Circuits **23**, 1171, (1998).
3. A. Bune, V. Fridkin, S. Ducharme, L. Blinov, S. Palto, A. V. Sorokin, S. Yudin, and A. Zlatkin, Nature **391**, 874 (1998).
4. F. Xia, B. Razavi, H. Xu, Z. Y. Cheng, and Q. M. Zhang, J. Appl. Phys. **92**, 3111 (2002).
5. N. Tsutsumi, A. Ueyasu, W. Sakai, and C. K. Chiang, Thin Solid Films, **483**, 340 (2005).
6. Q. M. Zhang, H. Xu, F. Fang, F. Xia, and H. You, J. Appl. Phys. **89**, 2613 (2001).
7. K. Tashiro, and R. Tanaka, Polymer, **47**, 5433 (2006).
8. A. A. Prabu, J. S. Lee, K. J. Kim, and H. S. Lee, Vib. Spec. **41**, 1 (2006).
9. J. S. Lee, K. J. Kim, and A. A. Prabu, Solid State Phenomena, **124-126**, 303 (2006).
10. J. S. Lee, A. A. Prabu, Y. M. Chang, and K. J. Kim, Macromol. Symp. **249-250**, 13 (2007).
11. A. A. Prabu, K. J. Kim, and C. Park, Vib. Spec. under revision (2008).
12. S. Fujisaki, H. Ishiwara, and Y. Fujisaki, Appl. Phys. Lett. **90**, 162902 (2007).

New Phase Change Memory
and Deposition Methods

Mater. Res. Soc. Symp. Proc. Vol. 1071 © 2008 Materials Research Society 1071-F09-09

Reversible Electrical Resistance Switching in GeSbTe Thin Films: An Electrolytic Approach without Amorphous-Crystalline Phase-Change

Ramanathaswamy Pandian, Bart J. Kooi, George Palasantzas, and Jeff Th. M. De Hosson
Department of Applied Physics, Zernike Institute for Advanced Materials, University of Groningen, Nijenborgh 4, Groningen, 9747 AG, Netherlands

ABSTRACT

Besides the well-known resistance switching originating from the amorphous-crystalline phase-change in GeSbTe thin films, we demonstrate another switching mechanism named 'polarity-dependent resistance (PDR) switching'. The electrical resistance of the film switches between a low- and high-state when the polarity of the applied electric field is reversed. This switching is not connected to the phase-change, as it only occurs in the crystalline phase of the film, but connected to the solid-state electrolytic behavior i.e. high ionic conductivity of (Sb-rich) GeSbTe under an electric field. I-V characteristics of nonoptimized capacitor-like prototype cells of various dimensions clearly exhibited the switching behavior when sweeping the voltage between +1 V and -1 V (starting point: 0 V). The switching was demonstrated also with voltage pulses of amplitudes down to 1 V and pulse widths down to 1 microsecond for several hundred of cycles with resistance contrasts up to 150 % between the resistance states. Conductive atomic force microscopy (CAFM) was used to examine PDR switching at nanoscales in tip-written crystalline marks, where the switching occurred for less than 1.5 V with more than three orders of resistance contrasts. Our experiments demonstrated a novel and technologically important switching mechanism, which consumes less power than the usual phase-change switching and provide opportunity to bring together the two resistance switching types (phase-change and PDR) in a single system to extend the applicability of GeSbTe materials.

INTRODUCTION

For the next generation nonvolatile memories, several random access memory (RAM) technologies have been proposed e.g. based on magneto-resistance (MRAM), ferroelectricity (FRAM), phase-change (PRAM) and electrical-resistance (RRAM). Among them, PRAM and RRAM based on *electrical resistance switching* have been given more focus in recent years as they prove to be promising candidates for the next generation nonvolatile memories. Being the active medium of the proposed phase-change and electrical-resistance based memories, chalcogenide materials are particularly promising and versatile. PRAM is based on resistance switching caused by amorphous-crystalline phase-change and is not voltage-polarity dependent. RRAM, on the other hand, is sensitive to polarity of the applied voltage and is not connected to phase-change. The resistance switching of RRAM is attributed to the electrolytic behavior and/or ionic conductivity of the material in the solid-state. Thus far, the phase-change and polarity-dependent resistance switching in chalcogenides were considered independently. For example, numerous compositions derived from the GeSbTe system showed amorphous-crystalline switching and Ag- or Cu-doped chalcogenides such as AgS, CuS, AgGeSe, AgGeTe and AgInSbTe showed PDR switching. In Ag/Cu-free chalcogenides, such as in GeSbTe system most commonly used in phase-change data storage, the PDR switching has not yet been demonstrated. AgInSbTe is the only material (comparable to GeSbTe system) for which the PDR switching is reported [1]. However, this was shown to have a high threshold voltage ($V_{th} > 10$ V), which is a significant drawback. In this paper, we demonstrate that PDR switching could be achieved with GeSbTe at various length scales using low voltages (≤ 2 V), which can not induce amorphous-crystalline phase transitions in the system.

EXPERIMENTS

The samples consist of 20 or 40 nm thick Sb-excess $Ge_2Sb_{2+x}Te_5$ (GST) phase-change films on 100 nm thick Mo bottom-electrodes on Si substrates. Mo and amorphous GST were deposited by dc-magnetron sputtering. Ag or Au was used as top-electrode of the capacitor-like prototype cells (figure 1a). With such memory cells, the PDR switching behavior was examined via I-V measurements using a Keithley 2601 source-meter with a voltage sweep rate of about 0.8 V/s. When the cell is on pulse mode operation, the dc voltage source is replaced by a pulse generator (Stanford Research System Inc., Model DG535) and the switching was tested for various pulse-amplitudes and pulse widths. To investigate the switching at nanoscales, conductive atomic force microscopy (CAFM, Veeco Dimension-3100) was used. Pt/Ir coated tip of the microscope served as a top-electrode in this case. The schematic of the CAFM experimental setup is shown in figure 1b. With the setup, in addition to topographs, current-images showing the local conductivity of the sample were obtained. Biasing the sample with smaller dc voltages ($<< V_{th}$ for phase-change) and measuring the electrical current (passing along the sample thickness) with a conductive AFM-tip that is virtually grounded, gives the conductance image. A high-gain current amplifier connected electrically in series with the tip detects currents down to 5 pA. In our experimental setup, scanning the amorphous area with lower dc bias voltages (1 to 2 V) did not show any significant current flow above this lower limit. For the PDR switching experiments, the volume of GST film underneath the top-electrode was electrically crystallized. Crystallization occurs due to the joule heating effect when current passing between the electrodes through GST film along its thickness. All the measurements were performed in air and at room temperature.

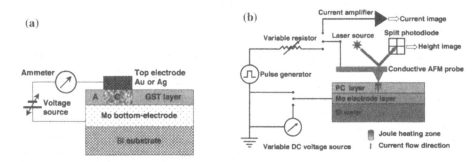

Figure 1. (a) Schematic (not to scale) of a capacitor-like cell structure with Ag or Al top-electrode; A and C refers to amorphous and polycrystalline phase of the GST layer, respectively. (b) CAFM experimental setup used for data recording studies at nanoscale. The setup facilitates both height- and current-imaging simultaneously.

RESULTS AND DISCUSSIONS

PDR switching with capacitor-like cells

To investigate PDR switching at macroscopic level (millimeter scale), capacitor-like memory cells with Ag top-electrodes were prepared (size of Ag electrodes was about 1 mm and thickness was ~ 0.1 mm). Figure 2a shows a typical I-V behavior of a such prototype cell. Note that before the measurements, the amorphous GST between the electrodes was crystallized by sweeping the applied voltage from zero to values > V_{th} for crystallization and PDR switching without this initial crystallization of GST turned out to be impossible. The sample is therefore initially at low-resistance state (LRS). When sweeping the voltage from zero to the negative values with respect to the bottom electrode, a linear I-V behavior is observed until the voltage reaches -0.4 V. Beyond this threshold voltage (V_{th} ≈ -0.4 V), the sample switches to a high-resistance state (HRS), which is ~ 5 times the LRS and remains at this state for further voltage sweep from -0.4 to -0.6 and from -0.6 to +0.4 in the other direction. An almost a linear I-V behavior is observed, at this HRS, for the voltage sweep from -0.6 to +0.4. Sweeping the voltage above +0.4 switches the sample back to LRS. The cutoff for large currents above 40 mA is due to a current limit set to prevent sample damage. I-V characteristic with the two different resistance states (see figure 2a) clearly exhibits the PDR switching behavior of the sample. This intrinsic memory effect was reproducible for a number of cycles within ±0.4 V. The two (meta-) stable resistance states are nonvolatile for several months as tested after switching and could be read with low bias voltages (<< V_{th}) of either polarity.

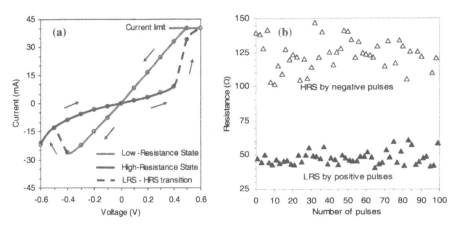

Figure 2. (a) Memory switching I-V behavior of a cell with Ag top-electrode. (b) LRS and HRS during the pulse-mode operation with voltage pulses of ±1.25 V and 1 μs.

Using voltage pulses for the switching includes the advantages of fast device operation and reducing thermal effects or damage. An example for the pulse-mode operation of a cell, for several number of cycles, exhibiting the two resistance states is shown in figure 2b. Voltage pulses of ±1.25 V and 1 μs were used to perform the switching. Negative pulses resulted LRS to

HRS switching, which is equivalent to a *reset* operation brings the sample to *off-state*. By positive pulses HRS to LRS transition occurs. This transition is equivalent to a *set* operation by which the sample returns to its *on-state*. Between each set and reset pulse, the resistance state of the sample was read with a voltage pulse of 0.1 V & 20 ms. The switching was reproducible and resistance states were stable for several months of testing. The contrast between the resistance states in this particular example is about 150 %. Capacitor-like memory cells with Aluminum top-electrodes of size ~ 150 µm and thickness of 1 µm were prepared to examine the switching at micron scale. I-V measurements showed the PDR switching behaviour and in the pulse-mode operation LRS-HRS transition was demonstrated (with voltage pulses of amplitudes down to 1 V and pulse widths down to 1 µs) for hundreds of cycles with resistance contrasts more than 100 % between the resistance states. Results with Al top-electrodes are not shown in this article, but they will be published elsewhere [2].

PDR switching studies at nanoscales using CAFM

Memory switching at nanoscales is essential to meet the ever growing demands in high data-storage densities. In recent years, a strong impetus in this direction has been given by AFM nanolithography [3]. Therefore, we explored the resistance switching at nanoscales by CAFM and in this case the conductive AFM-tip with about 50 nm radius of curvature, served as top electrode to the GST film. Figure 3a is a contact-mode AFM topograph showing an array of tip-written crystalline marks in an amorphous GST film. These marks were written by injecting dc voltage pulses of amplitude -5 V and width 500 ms from the tip into the electrically grounded film. The written marks are visible as pits at nanoscale because of the (local) density reduction of the amorphous film upon crystallization under the tip. One of the crystalline marks was taken to examine the PDR switching at nanoscale. Continuous scanning of an area including the mark with a positive bias of 1.5 V (i.e. the film is biased with respect to the virtually grounded tip), referred as a set operation, brought it into a LRS or on-state. On the other hand, a negative bias scanning (-1.5 V), referred as a reset operation, takes the mark to a HRS or off-state. When the set operation is repeatedly performed with +1.5 V, the mark switches back to its LRS or on-state. Simultaneous measurement of current flow through the tip during scanning allowed mapping the resistance state of the mark during the set/reset operations. The background amorphous phase resistance during these operations remains below the current detection limit (i.e. 5 pA) of our CAFM setup. Figures 3b, 3c & 3d are topographs of the mark during the set/reset operations. Figures 3e, 3f & 3g are current-images showing the LRS, HRS and LRS of the mark during the *set*, *reset* and *set* operations, respectively. Figures 3h, 3i & 3j are current-profiles corresponding to the current-images 3e, 3f and 3g, respectively.

Note that the HRS of the mark is electrically indistinguishable from the surrounding amorphous phase (cf. 3f & 3i), because the current flow across both is lower than 5 pA and thus lower than the detection limit. During the set/reset operations topography of the mark did not alter markedly (cf. 3b, 3c & 3d), indicating that there is no structural change involved with the PDR switching. The ON-state current profiles, 3h & 3j, reveal that the mark has a resistance contrast of more than 3 orders of magnitude with its surrounding. The operating voltage of this switching (±1.5 V) is clearly lower than the threshold voltage (> ±4 V) required for the phase transition. This indicates that the PDR type switching is more advantageous for future device applications. Since the switching causes no detectable density changes, it should also be advantageous from the cyclability point of view. In current images, 3e & 3g, it can be seen that the electrical conductivity (or resistivity) of the crystalline mark (within the dashed circle) is not

184

uniform. A considerable area fraction is still at HRS, where it should be homogeneous at LRS. This can be due to several factors, e.g. an incomplete crystallization by the tip, an improper tip-sample electrical contact due to relative fast tip scanning, surface roughness of the sample and removal of the conductive coating from the tip. Our experiments also prove that the switching is not limited by the switching area or electrode material type, but the switching speed is limited by a few experimental constraints i.e. AFM scanning speed and pulse-shape loss due to the large capacitance of our non-optimized test structures (for pulse durations < a few hundred μs).

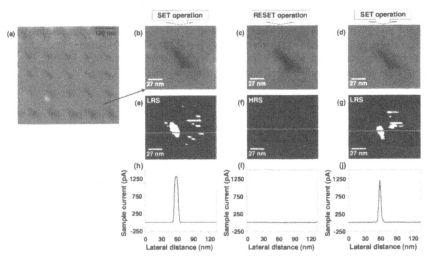

Figure 3. (a) AFM topograph showing an array of tip-written crystalline marks in a 40 nm amorphous GST film. (b), (c) & (d) are topographs of a crystalline mark during the set, reset & set operations, respectively. (e), (f) & (g) are current-images recorded with biasing of ±1.5 V showing on-, off- and on-states of the mark, respectively. (h), (i) & (j) are on-, off- and on-state current profiles, corresponding to the current-images (e), (f) & (g), respectively.

Proposed mechanism of PDR switching

Resistance switching driven by the polarity of the applied electric field can be related to the solid-state electrolytic behavior of the chalcogenide material. When the material is subjected to an electric field, electrochemical reactions near the electrodes lead to ionic conduction. If the electric field is sufficiently strong, electrically conductive filamentary pathways form between the electrodes leading to an LRS and if the polarity is reversed, the pre-existing paths become discontinuous due to ion migration in the opposite direction resulting in a HRS. In previous reports, chalcogenides showing this behavior include AgS [4], CuS [5], AgGeSe [6], AgGeTe [7] and AgInSbTe [8], where the LRS and HRS are a result of the formation and rupture of Ag- or Cu-filaments. A similar electrolytic mechanism can explain the PDR switching in our Sb-excess GeSbTe system. In this case, conductive Sb instead of Ag- or Cu-filaments can be formed and dissolved in amorphous phase that still persists with a small volume fraction when the GST crystallites are formed.

Points favoring of this mechanism are: (1) crystallization of this type of material leads to phase separation, where the stoichiometric nanosized $Ge_2Sb_2Te_5$ crystals form with the excess Sb as amorphous phase at the grain boundaries [9] and (2) cross-sectional transmission electron microscopy (TEM) studies showed that in GeSbTe films a strong tendency exists to form crystallites near the film surface leaving some amorphous volume near the film-substrate interface [10]. Note that (metallic) Sb is several orders of magnitude more conductive than Ge or Te within GST system. Therefore, when a sufficiently strong electric field is applied, conducting dendrite-like Sb-filaments form and can bridge the $Ge_2Sb_2Te_5$ grains through the amorphous matrix with the electrodes. The conducting Sb-bridges persist until they are dissolved/ruptured by reversing the polarity of the electric field. Instead of Sb-filaments also a similar (electrolytic) mechanism where grain-boundaries can switch between a conductive and insulating state can explain the observations. Cross-section TEM studies are in progress to investigate the filament formation or the conductive or insulating grain boundaries in this GST system.

CONCLUSIONS

We demonstrated the existence of polarity-dependent resistance switching in crystallized Sb-excess $Ge_2Sb_2Te_5$ films at various length scales. In prototype capacitor-like cells of size down to micrometer, voltage pulses showed this switching within time scales of micro-seconds with resistance contrasts up to 150 % between the resistance states. Moreover, using conductive atomic force microscopy, switching was also possible at nanoscales with a better resistance contrast of more than three orders of magnitude. Notably, the operating voltage of the write/erase process ($< \pm1.5$ V) is much lower compared to that of the current ferroelectric and flash memories and is also compatible with future microelectronic and data-storage systems. Our study shows conclusively that in a single material phase-change and polarity-dependent switching can be combined from macroscopic down to nanoscales.

REFERENCES

1. Y. Yin, H. Sone and S. Hosaka, Jpn. J. Appl. Phys. 45, 4951 (2006).
2. R. Pandian, B. J. Kooi, G. Palasantzas, J. T. M. De Hosson and A. Pauza, In preparation.
3. S. Gidon, O. Lemonnier, B. Rolland, O. Bichet, C. Dressler and Y. Samson, Appl. Phys. Lett. 85, 6392 (2004); K. Tanaka, J. Non-Cryst. Solids 353, 1899 (2007); C. D. Wright, M. Armand and M. M. Aziz, J. Hist. Astron. 5, 50 (2006).
4. Y. Hirose and H. Hirose, J. Appl. Phys. 47, 2767 (1976); K. Terabe, T. Hasegawa, T. Nakayama and M. Aono, Nature (London) 433, 47 (2005).
5. T. Sakamoto, NEC J. Adv. Tech. 2, 260 (2005).
6. M. N. Kozicki, M. Park and M. Mitkova, IEEE Trans. Nanotechnol. 4, 331 (2005).
7. C.-J. Kim and S.-G Yoon, J. Vac. Sci. Technol. B 24, 721 (2006).
8. Y. Yin, H. Sone and S. Hosaka, Jpn. J. Appl. Phys., Part 1 45, 4951 (2006).
9. N. Yamada and T. Matsunaga, J. Appl. Phys. 88, 7020 (2000).
10. S.-M. Yoon, K.-J. Choi, N.-Y. Lee, S.-Y. Lee, Y.-S. Park and B.-G. Yu, Jpn. J. Appl. Phys., Part 2 46, L99 (2007); J. A. Kalb, C. Y. Wen, F. Spaepen, H. Dieker and M. Wuttig, J. Appl. Phys. 98, 054902 (2005); T. H. Jeong, M. R. Kim and H. Seo, ibid. 86, 774 (1999).

Mater. Res. Soc. Symp. Proc. Vol. 1071 © 2008 Materials Research Society　　　　　1071-F09-10

CVD of Amorphous GeTe Thin Films

Philip S Chen, William J Hunks, Matthias Stender, Tianniu Chen, Gregory T Stauf, Chongying Xu, and Jeffrey F Roeder
ATMI, Danbury, CT, 06810

ABSTRACT

Fourteen germanium (Ge) and two tellurium (Te) precursors were used to deposit Ge and Te thin films by a thermal MOCVD process using various co-reactant gases on TiN/Si and SiO_2/Si substrates. Selected results are presented in this paper. Smooth amorphous GeTe films with Te content as high as 48 at% were deposited. Annealing of the amorphous GeTe films at 400 °C under N_2 yielded smooth crystalline films that displayed a phase-change induced electrical resistivity reduction of over 300 times.

INTRODUCTION

Chalcogenide $Ge_2Sb_2Te_5$ (GST225) based phase-change memories (PCM) are one of the most promising candidates for next-generation non-volatile memories, with the potential to improve the performance over Flash memories as well as scale beyond the limits of current Flash technology. Most reports of GST films utilize sputtering for deposition in PCM devices. Although functional devices can be produced by this method, it is mostly suitable for planar devices or structures with relaxed geometric features. For the case of planar structures, high reset currents represent a significant challenge. For continued scaling of nano-electric device architectures, and for improved performance and lower costs, deposition into more challenging non-planar structure is needed. Reduction of reset current from 500 μA levels typical of planar devices to below 260 μA using a confined cell structure has been demonstrated by limiting the switched volume of the GST material to the contact plug.[1,2] This also greatly reduces the heat dissipation to the surrounding materials. Therefore, it is necessary to deposit the GST films using a process that offers good conformality such as ALD or CVD for higher aspect ratio structures in high density memory devices. In this paper, we report CVD of GeTe films from several classes of metalorganic precursors.

EXPERIMENT

A load locked single wafer deposition chamber equipped with a liquid delivery system and a vaporizer was used for the deposition process. Ge precursors, developed and purified at ATMI, were dissolved in solvent and delivered through a liquid mass flow controller into the vaporizer, where they were mixed with helium carrier gas and introduced through a temperature controlled showerhead above the wafer. A room temperature bubbler was used to provide the alkyl-telluride precursors for the GeTe deposition. Deposition experiments were carried out at chamber pressures between 0.9 and 8 Torr and wafer temperatures between 200 and 440 °C.

Both thermally oxidized Si and 50 nm TiN coated Si wafers were used for Ge and GeTe thin film depositions. The film composition was characterized with XRF, EDS and RBS analyses, while the microstructure was observed by scanning electron microscopy (SEM). The film thickness was measured by XRF using a sputtered $Ge_2Sb_2Te_5$ standard and verified with cross-sectional SEM, and the crystallinity and structure of the Ge and GeTe films were determined by XRD using a Rigaku diffractometer with excited CuKα radiation. A CDE-273 ResMap was used to determine the sheet resistance of the GeTe films on SiO_2 substrates.

DISCUSSION

Ge Deposition

Ge precursors with high deposition rates at low-temperature are needed for the deposition of amorphous GeTe, as high temperature deposition tends to form crystalline films with rough surface morphologies. In order to achieve this, several precursors were explored.

Ge Amide precursors

An Arrhenius plot of a Ge amide deposition onto TiN at pressures of 0.9 and 8 Torr is shown in figure 1. The deposition rates dropped precipitously between 280 and 360 °C, depending on the pressure and co-reactant gas used. The strong step functions in deposition rate are attributed to poor nucleation on the TiN surface, presumably due to an oxide surface layer.

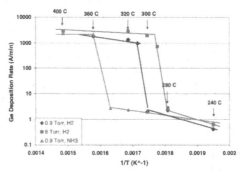

Figure 1. Arrhenius plot of a germanium amide precursor at 0.9 and 8 Torr pressure.

Ge Cyclopentadienyl precursors

Figure 2 shows an Arrhenius plot of Ge deposition from a cyclopentadienyl (Cp) type Ge precursor at 8 Torr pressure. Mass transport limited deposition occurs at temperatures greater than 300 °C. We also observed little difference in CVD behavior for both H_2 and NH_3 co-reactants.

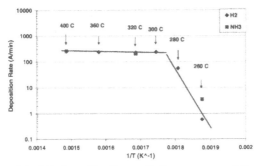

Figure 2. Arrhenius plot of Ge deposition from a GeCp precursor at 8 Torr pressure.

Ge Amidinate precursors

Arrhenius plots of Ge deposition from two Ge amidinate precursors at a deposition pressure of 8 Torr are illustrated in figure 3. The deposition rates of Ge with ammonia co-reactant were about two orders of magnitude higher than those with hydrogen co-reactant at around 300 °C. The significant differences in deposition rates between H_2 and NH_3 co-reactant allows an ALD-like deposition process with this precursor. Mass transport limited depositions were observed with amidinate 2 with deposition temperature greater than 280 °C.

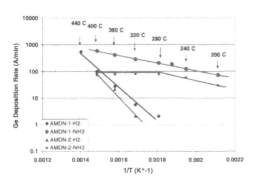

Figure 3. Arrhenius plots of Ge deposition from two amidinate types of Ge precursors at 8 Torr pressure. Ammonia co-reactant improves the deposition rate significantly over hydrogen co-reactant.

Ge Guanidinate precursors

Figure 4 shows the Arrhenius plot of Ge deposition from a guanidinate type of Ge precursor at 8 Torr pressure. Mass transport limited depositions were detected for both H_2 and

NH_3 co-reactants at temperatures above 320 and 280 °C, respectively. The deposition rates with H_2 co-reactant are about 10 times higher than those with NH_3 co-reactant.

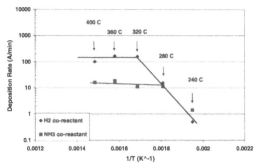

Figure 4. Arrhenius plot of Ge deposition from a guanidinate type of Ge precursor at 8 Torr pressure.

Te Deposition

Figure 5 shows the Arrhenius plot of Te deposition behavior from two alkyl tellurides at a pressure of 8 Torr with H_2 and NH_3 co-reactants. No significant differences in deposition rate were detected between H_2 and NH_3 co-reactants. Di-isopropyl telluride yielded a significantly higher deposition rate than di-tert-butyl telluride.

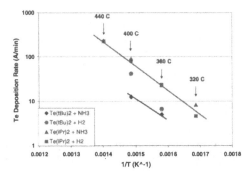

Figure 5. Arrhenius plot of Te deposition from two alkyl telluride precursors at a pressure of 8 Torr.

GeTe Deposition

The crystallinity of the as-deposited GeTe thin film depended strongly on deposition temperature and pressure, with amorphous film deposited at lower temperature. We observed that the stoichiometric GeTe films were all crystalline, while the Te lean films were mostly amorphous. Figure 6 shows a highly conformal amorphous GeTe films deposited on a 0.18 μm trench structure. The Te content of this amorphous GeTe film was determined to be about 40 at%.

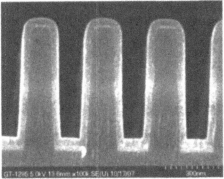

Figure 6. Highly conformal amorphous GeTe deposition on a 0.18 μm trench structure.

X-ray diffraction patterns of an as-deposited and annealed GeTe film (with 40 at% Te) on TiN substrate are displayed in figure 7. Annealing of this amorphous GeTe film at 400 °C under N₂ yielded a smooth crystalline film, as shown in figure 8. Annealing of an amorphous GeTe film with about 20 at% Te on a SiO₂ substrate at 400 °C reduced the film sheet resistance from over 5 MΩ/square to about 15 kΩ/square.

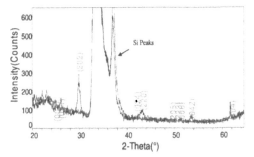

Figure 7. XRD patterns of a GeTe film with 40 at% Te, before and after 400 °C, N₂ anneal.

Figure 8. Surface morphology of the smooth GeTe films, as-deposited and post-annealed.

CONCLUSIONS

Fourteen germanium and two alkyl telluride precursors were investigated for the CVD of Ge and Te thin films. Reasonable deposition rates of Ge or Te film can be achieved over a temperature range between 200 and 400 °C. Smooth amorphous GeTe films with Te content consistently between 40 and 50 at% Te and with good conformality on a 0.18 μm trench structure has been demonstrated. Annealing of this amorphous GeTe film at 400 °C under N_2 yielded a smooth crystalline film with sheet resistance reduced by greater than two orders of magnitude.

ACKNOWLEDGMENTS

The authors would like to thank Drs. Weimin Li and Jun-Fei Zheng for helpful discussions and critical review of the manuscript.

REFERENCES

1. J. I. Lee, H. Park, S. L. Cho, Y. L. Park, B. J. Bae, J. H. Park, J. S. Park, H. G. An, J. S. Bae, D. H. Ahn, Y. T. Kim, H. Horii, S. A. Song, J. C. Shin, S. O. Park, H. S. Kim, U-In. Chung, J. T. Moon, and B. I. Ryu, *2007 Symposium of VLSI Technology Digest of Technical Papers*, 6B-4.
2. M. Breitwisch, T. Nirschl, C.F. Chen, Y. Zhu, M.H. Lee, M. Lamorey, G.W. Burr, E. Joseph, A. Schrott, J.B. Philipp, R. Cheek, T.D. Happ, S.H. Chen, S. Zaidi, P. Flaitz, J. Bruley, R. Dasaka, B. Rajendran, S. Rossnagel, M. Yang, Y.C. Chen, R. Bergmann, H.L. Lung, and C. Lam, *2007 Symposium of VLSI Technology Digest of Technical Papers*, 6B-3.

Mater. Res. Soc. Symp. Proc. Vol. 1071 © 2008 Materials Research Society 1071-F09-11

Germanium ALD/CVD Precursors for Deposition of Ge/GeTe Films

William Hunks, Philip S. Chen, Tianniu Chen, Matthias Stender, Gregory T. Stauf, Leah Maylott, Chongying Xu, and Jeffrey F. Roeder

ATMI, 7 Commerce Dr., Danbury, CT, 06810

ABSTRACT

In order to deposit conformal films in the high aspect ratio trench and via structures in future high-density phase-change memory devices, suitable ALD/CVD precursors are needed. We report on the development of novel germanium(II) metal-organic ALD/CVD precursors containing amide, cyclopentadienyl, and amidinate ligands. The physical properties, volatility, and thermal behavior of the precursors were evaluated by simultaneous thermal analysis (STA) and vapor pressure measurements. Stability studies were conducted to investigate the suitability of the precursors for use as ALD/CVD precursors for device manufacturing.

INTRODUCTION

As device manufacturers continue to increase storage capacity by scaling to smaller dimensions, it is anticipated that flash memory will experience performance limitations due to its use of electrons to store data. Several groups are investigating phase-change random access memory (PRAM) as an alternative memory technology that is scalable to < 5nm and compatible with Si-based ICs [1-2]. PRAM combines the fast memory access speed of DRAM/SRAM with the non-volatile storage feature of flash. PRAM displays long-term data retention at elevated temperatures, fast read/write speeds of 5-20ns, and is capable of direct overwrites of up to 10^8 cycles. Digital data is stored through a thermally reversible phase transition from an amorphous state (high resistivity, binary 0) to a crystalline state (low resistivity, binary 1) with fast crystallization kinetics.

For prototype PRAM devices, GST (germanium-antimony-tellurium) has emerged from the chalcogenide glasses used in the manufacturing of rewritable optical disks. However, for the high density memory arrays proposed for PRAM devices of less than 50nm, the increasing aspect ratio makes filling of the trenches and vias difficult. For conformal deposition without voids, there is a strong need to transition from physical vapor deposition (PVD) to chemical vapor deposition (CVD), which requires suitable low-temperature deposition precursors.

CVD precursors explored use volatile Ge(IV) compounds that require high temperatures for deposition and give poor film conformality and morphology. Reported GST films have been deposited using CVD processes at temperatures ranging from 300-400°C using tetraallylgermane or tetraisobutylgermane co-deposited with triisopropylantimony and diisopropyltellurium. Deposited films displayed nonuniform and island growth behavior [3-7]. Similarly, co-deposition of $Ge(NMe_2)_4$, $Sb(NMe_2)_3$, and iPr_2Te gave poor film morphology [8]. We describe a series of Ge(II) precursors that show enhanced low-temperature deposition characteristics, investigate their thermal stability, vapor pressure, and thin film deposition rates and contrast them to the Ge(IV) precursor $Ge(NEtMe)_4$.

EXPERIMENT

CVD precursors were synthesized and purified at ATMI. Thermogravimetric analysis (TGA) and differential scanning calorimetry (DSC) were collected on a Netzsch STA 449C instrument under an inert argon atmosphere. Deposition studies were conducted using a load-locked single wafer warm-walled reactor. Delivery lines and chamber walls were maintained at approximately 100°C and a chamber pressure of 0.9 to 8 Torr during deposition experiments. Precursors were delivered into the CVD chamber using a nitrogen carrier gas and either a H_2 or NH_3 co-reactant gas. Ge and GeTe films were deposited onto cleaved sections of 50 nm TiN coated Si wafers at various substrate temperatures. Film composition was characterized by wavelength dispersive X-Ray Fluorescence (XRF) and EDS analysis. All vapor pressure measurements were collected by the Knudsen cell effusion method.

DISCUSSION

In order to use MOCVD precursors in deposition experiments, a number of criteria must be met, including high volatility, thermal stability, and adequate growth rates for the deposition of Ge-based PRAM alloys in sub-50nm trench and via structures. Unfortunately, these parameters are often in contrast to the low-temperature deposition processes needed to facilitate conformal deposition. Thermally unstable metal-ligand complexes suitable for the low-temperature CVD of Ge and GeTe films lead to precursors with reduced vapor pressure making standard bubbler delivery methods difficult without causing decomposition of the precursors. A series of divalent germanium compounds were investigated and shown to have sufficient stability and volatility for precursor delivery into the deposition chamber under standard bubbler temperature and pressure conditions.

Table I. STA data on germanium precursors.

Precursor	m.p. (°C)	T_{50} (°C)	Mass residue (%)
$Ge(NEtMe)_4$	liq. at RT	148	0.0
$Ge\{N(SiMe_3)_2\}_2$	32	175	5.4
$Ge\{N(SiMe_3)(tBu)\}_2$	liq. at RT	184	9.5*
$Ge(C_5Me_5)_2$	83	227	3.8
$Ge(C_5Me_4Pr)_2$	liq. at RT	215	9.7
$Ge\{MeC(iPrN)_2\}_2$	54	211	16.4

* mass residue was measured at 210°C.

Thermal stability studies

The volatility and thermal stability of the precursors were investigated by thermogravimetric analysis (TGA) and are summarized in table I. The melting points were obtained from the endothermic transition in the DSC curve during STA analysis. The precursors studied are either liquids, or low melting point solids that are liquids at typical bubbler temperatures leading to controllable vapor pressure and delivery flux. Selected TGA curves are presented in figure 1. Relative volatilities were based on the T_{50}, which is the temperature where half of the precursor has evaporated. Based on T_{50} results, the germanium(II) amides [Ge{N(SiMe$_3$)$_2$}$_2$, and Ge{N(SiMe$_3$)(tBu)}$_2$] are more volatile than the Ge(II) cyclopentadienyl or amidinate complexes. The Ge(II) amides and cyclopentadienyls leave nonvolatile residues during transport of 5-10%. However, the complexes display a single sharp transport step in the TGA curve with little mass residue indicating they have sufficient thermal stability for vaporization without significant decomposition. On the other hand, the amidinate complex Ge{MeC(iPrN)$_2$}$_2$, yielded a multi-step transport TGA curve with corresponding endothermic peaks in the DSC plot that suggests thermal decomposition. All the Ge(II) complexes studied are somewhat less volatile than the Ge(IV) compound Ge(NEtMe)$_4$.

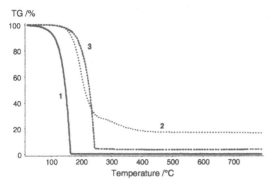

Figure 1. TGA curves for the precursors: Ge(NEtMe)$_4$, 1; Ge{MeC(iPrN)$_2$}$_2$, 2; Ge(C$_5$Me$_5$)$_2$, 3.

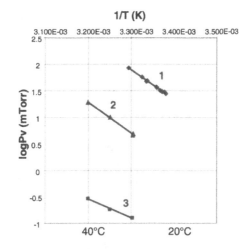

Figure 2. Arrhenius plot of vapor pressure for the precursors Ge(NEtMe)$_4$, 1; Ge{N(SiMe$_3$)(tBu)}$_2$, 2; and (C$_5$Me$_4$Pr)$_2$Ge, 3.

Vapor pressure measurements

The vapor pressure for the series of compounds: Ge(NEtMe)$_4$, Ge{N(SiMe$_3$)(tBu)}$_2$, and (C$_5$Me$_4$Pr)$_2$Ge, were measured via the Knudsen cell effusion method. The Arrhenius plots of logP(mTorr) against reciprocal temperature are displayed in figure 2. At least three points were collected for each precursor, and then reproduced, indicating that no decomposition occurred during the sampling measurement. The resulting vapor pressures and enthalpies of vaporization calculated from the measured data are summarized in table II. The vapor pressure follows the trend of Ge(NEtMe)$_4$ > Ge{N(SiMe$_3$)(tBu)}$_2$ > (C$_5$Me$_4$Pr)$_2$Ge. This verifies the T$_{50}$ results showing the Ge(II) amide is more volatile than either the Ge(II) amidinate or cyclopentadienyl complexes, although less volatile than Ge(NEtMe)$_4$. Extrapolation of the vapor pressure curve for the least volatile precursor Ge(C$_5$Me$_4$Pr)$_2$, gives a vapor pressure of 0.06 mTorr at 20°C that increases to 347 mTorr at 150°C. Nonetheless, this precursor is readily distillable at 150°C under vacuum indicating even the least volatile precursor described has sufficient volatility and thermal stability for transport into CVD chambers.

Table II. Vapor pressure measurements of germanium precursors.

Precursor	ΔHvap (KJ/mol)	Vapor pressure {logP(mTorr)}
Ge(NEtMe)$_4$	105.3	-5500.5/T (K) + 20.045
Ge{N(SiMe$_3$)(tBu)}$_2$	110.3	-5763/T (K) + 19.723
Ge(C$_5$Me$_4$Pr)$_2$	68.4	-3574/T (K) + 10.897

Table III. Deposition rates of germanium CVD precursors.

Precursor	Co-reactant gas	Temperature (°C)	Ge deposition rate (Å/min)
Ge(NMe$_2$)$_4$	H$_2$	240	< 1
	H$_2$	320	2940
Ge{N(SiMe$_3$)$_2$}$_2$	NH$_3$	240	< 1
	NH$_3$	300	7.98
Ge{N(SiMe$_3$)(tBu)}$_2$	NH$_3$	240	1.01
	NH$_3$	320	1.22
Ge{MeC(iPrN)$_2$}$_2$	NH$_3$	240	123.5
	NH$_3$	320	282.7
Ge(C$_5$Me$_5$)$_2$	H$_2$	240	< 1
	H$_2$	320	220

Deposition experiments

Deposition experiments were conducted using the precursors: Ge(NMe$_2$)$_4$, Ge[N(SiMe$_3$)$_2$]$_2$, Ge[N(SiMe$_3$)(tBu)]$_2$, Ge{MeC(iPrN)$_2$}$_2$, and (C$_5$Me$_5$)$_2$Ge and are listed in table III. A thermal MOCVD process was used to grow metallic germanium films using NH$_3$ or H$_2$ co-reactant gases on TiN/Si substrates or SiO$_2$/Si substrates. The germanium(IV) amide, Ge(NMe$_2$)$_4$, shows a rapid deposition rate at 320°C under H$_2$, however at 240°C no significant film growth was observed (<1 A/min). CVD experiments using the germanium(II) amides Ge{N(SiMe$_3$)$_2$}$_2$, Ge{N(SiMe$_3$)(tBu)}$_2$, and Ge{N(SiMe$_2$CH$_2$CH$_2$SiMe$_2$)}$_2$ was recently reported [9]. These germanium(II) amides show minimal Ge film growth rates at 240 and 320°C, with a maximum growth rate of 8Å/min obtained for Ge{N(SiMe$_3$)$_2$}$_2$. The Ge(II) cyclopentadienyl complex, (C$_5$Me$_5$)$_2$Ge, exhibits a high Ge film growth rate independent of the H$_2$/NH$_3$ co-reactant gas (~200Å/min). However, a slow deposition rate was observed at 240°C (1-2 Å/min). The Ge(II) amidinate, Ge{MeC(iPrN)$_2$}$_2$, gave a high deposition rate down to 240°C (123 Å/min) under NH$_3$.

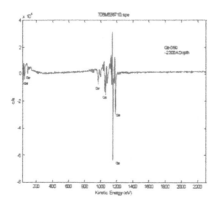

Figure 3. Auger spectrum of a 0.8μm Ge film deposited with Ge(NMe$_2$)$_4$ showing < 2% N, C contamination.

From the vapor pressure studies, the Ge(II) cyclopentadienyl and amidinate complexes are less volatile than either the Ge(IV) amide or Ge(II) amides studied. However, these precursors show increased deposition rates at lower temperatures which is required for the manufacturing of high-density memory chips. Thermal stability studies indicate that Ge(II) amidinates and cyclopentadienyls have sufficient stability to be delivered into CVD chambers with minimal decomposition. Although, the deposition rates of Ge(II) amides are somewhat slower, they may be desired in situations where composition, conformality, and smooth morphologies are necessary. Furthermore, we were able to deposit Ge$_x$Te$_{1-x}$ alloys with iPr$_2$Te.

CONCLUSIONS

This study investigated several Ge (IV) and Ge (II) chemistries for their use as precursors for CVD/ALD of GST. It was shown that a high deposition rate of Ge can be achieved in a wide temperature window of 200 to 400°C. Germanium (II) precursors with cyclopentadienyl or amidinate ligands have enhanced deposition rates at low-temperatures relative to Ge(NEtMe)$_4$. These germanium compounds have sufficient volatility, thermal stability, and Ge deposition rates indicating they are viable precursors for MOCVD of Ge and GeSbTe films.

ACKNOWLEDGMENTS

The authors would like to thank Drs. Weimin Li and Jun-Fei Zheng for helpful discussions and critical review of the manuscript.

REFERENCES

1. M. Wuttig, N. Yamada, *Nature Mat.* **6**, 824 (2007).
2. S. Hudgens, B. Johnson, *MRS Bulletin* **29**, 829 (2004).
3. R.Y. Kim, H.G. Kim, S.G. Yoon, *Appl. Phys. Lett.* **89**, 102107 (2006).
4. B.J. Choi, S. Choi, Y.C. Shin, C.S. Hwang, J.W. Lee, J. Jeong, Y.J. Kim, S.Y. Hwang, S.K. Hong, *J. Electrochem. Soc.* **154**(4), H318 (2007).
5. B.J. Choi, S. Choi, Y.C. Shin, K.M. Kim, C.S. Hwang, Y.J. Kim, Y.J. Son, S.K. Hong, *Chem. Mater.* **19**, 4387 (2007).
6. R.Y. Yim, H.G. Kim, S.G. Yoon, *J. Appl. Phys.* **102**, 083531 (2007).
7. D.V. Shenai, R.L. DiCarlo Jr., M.B. Power, A. Amamchyan, R.J. Goyette, E. Woelk, *J. Cryst. Growth* **298**, 172 (2007).
8. J. Lee, S. Choi, C. Lee, Y. Kang, D. Kim, *Appl. Surf. Sci.* **253**, 3969 (2007).
9. T. Chen, C. Xu, W. Hunks, M. Stender, G.T. Stauf, P.S. Chen, J.F. Roeder, *ECS Transactions* **11**(7), 269 (2007).

Future Explorative Memory
Concepts

Mater. Res. Soc. Symp. Proc. Vol. 1071 © 2008 Materials Research Society 1071-F01-07

Concentric Metallic-Piezoelectric Microtube Arrays

H.J. Fan[1], S. Kawasaki[1], J. M. Gregg[2], A. Langner[3], T. Leedham[4], and J. F. Scott[1]
[1]Department of Earth Sciences, University of Cambridge, Cambridge, CB2 3EQ, United Kingdom
[2]Centre for Nanostructured Media, Queen's University of Belfast, Belfast, BT7 1NN, United Kingdom
[3]Max Planck Institute of Microstructure Physics, Halle, 06120, Germany
[4]Multivalent Ltd., The Laboratory, Eriswell, IP27 9BJ, United Kingdom

ABSTRACT

Trilayer concentric metallic-piezoelectric-metallic microtubes are fabricated by infiltrating porous Si templates with sol precursors. $LaNiO_3$ (LNO) is used as the inner and outer electrode material and $PbZrTiO_3$ (PZT) is the middle piezoelectric layer. Structure of the microtubes is characterized in details using scanning and transmission electron microscopy which are equipped with energy dispersive x-ray spectroscopy for elemental mapping. The hysteresis of a trilayer thin film structure of Pt-PZT-LNO is shown. These trilayer microtubes might find applications in inkjet printing.

INTRODUCTION

The infiltration technique utilizing porous templates appears to be a generic method for fabrication of nano and microtubes of a large variety of materials including metals and dielectric oxides (1,2) and metals (3,4). A number of reports have been made on fabrication of ferroelectric/piezoelectric nano- and microtubes through infiltration of trenched templates (3-8). Other techniques have also merged, e.g., $Pb(Zr_xTi_{1-x})O_3$ (PZT) and $BaTiO_3$ nanotubes by physical coating of vertical nanowires (9) and carbon nanotubes (10), quartz tubes by Li-catalyzed oxidation of porous Si (11), $PbTiO_3$ microtubes by chemical reaction of pre-grown TiO_2 nanotubes with PbO vapor (12).

For the application as piezoelectricity-based nanofluidic channels (e.g., liquid spray), a ferroelectric tube with a tube diameter in the micrometer range is more suitable than nanotubes. For example, the nozzle size of an inkjet printer head or a liquid drug sprayer is tens of micrometers, therefore, arrays of the piezotubes with an inner tube diameter of 1-10 micron are needed. In this content, the infiltration technique is a simple and cheap method compared to MOCVD (13), sputtering, and atomic layer deposition. An array of electrode-piezoelectric-electrode concentric tubes should be possible by multiple infiltration of the porous template. Bharadwaja et al.(8) conducted a systematic synthesis and structural characterization work on PZT microtubes sandwiched by $LaNiO_3$ and Pd concentric layers. They applied a vacuum-assisted infiltration which appears to be advantageous over the ambient-pressure infiltration and showed a linear increase of the tube wall thickness with increasing the number of infiltrations.

The property characterization was made either azimuthally by scanning a conductive AFM tip over the surface of the tubes which lie horizontally on a platinized substrate (9), or longitudinally from two ends of the tubes (6). No work has been done so far to measure the radial ferro/piezo properties from inner to outer sides of the tubes. This is a challenge because it requires a continuous coating of the tubes with electrode layers forming concentric trilayer structure, and the probe (an AFM tip or metal tip inside SEM) must go inside and touch the inner surface of a tube. The trilayered 3-D capacitor structures fabricated by Bharadwaja et al. (8), Alex et al. (9) and Kawasaki et al. (10) make a capacitance-voltage measurement of possible, although not demonstrated yet.

We are continuing the work initiated by Bharadwaja et al. (8), i.e., consequential infiltration of microporous silicon with LNO, PZT, LNO sol solutions towards a concentric structured metallic-piezoelectric-metallic trilayer structure. Several important issues must be addressed, like interface reaction problem, individual addressing, and tube wall homogeneity. Hereby we show our preliminary results of the structure and phase of the tubes.

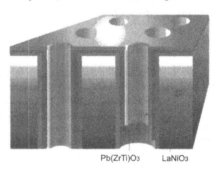

Pb(ZrTi)O3   LaNiO3

Figure 1 Schematics of the concentric microtubes embedded within a Si medium.

EXPERIMENTS

The porous Si templates were fabricated in MPI-Halle by etching of n-type phosphorus doped Si wafer. The doping density is dependent on the desired pore size. For a sample with 2 μm pore diameter, the resistivity is 5 Ω·cm, corresponds to a doping density of 1×10^{15} cm^{-3}. Due to the characteristics of the etching process, the ratio of the inter-pore distance to the pore diameter is nearly constant at 2. The pore depth is 50 μm.

The $Pb(Zr_{0.4}Ti_{0.6})O_3$ precursor is a metalorganic decomposition (MOD)-type solution provided by Kojundo chemical laboratory. The composition of the solution in the ratio of Pb/Zr/Ti was 1.1/0.4/0.6. The initial concentration of 17 wt% was diluted to 4 wt% by adding methyl ethyl ketone. The $LaNiO_3$ sol precursor was synthesized from nickel acetate $[Ni(CH_3COO)_2 \cdot 4H_2O]$ and lanthanum nitrate $[La(NO_3)_3]$. Briefly, nickel acetate was dissolved in acetic acid and an equimolar amount of lanthanum nitrate was dissolved in distilled water, both at room temperature. The two solutions were then mixed together with constant stirring. Dimethylformamide was added to the solution in order to avoid crack formation during heating. The final concentration of the precursor solution was 0.3 M.

For each deposition, the porous Si template was immersed into the solution and a slight vacuum was applied above the solution. The sample was then dried at 80 °C for about 30 min, then heated at 300 °C for another 30 min, followed by annealing in a resistant furnace at 650 °C (in case of PZT) or 700 °C (in case of LNO) for ca 20 min. The infiltration process was repeated twice for both LNO and PZT.

Before characterization, the tube arrays were exposed from the Si matrix. The samples were first polished to remove the surface LNO/PZT layer (which would otherwise prevent the Si etching), and then dipped in a 30 wt% KOH solution at room temperature for about 10 min. In order to have tube ends open, the samples were sonicated for several seconds after the polishing, and then back into the KOH solution for ca 2 hr. The samples were characterized using SEM, TEM and XRD.

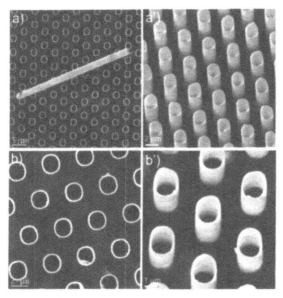

Figure 2 SEM images of the microtube arrays of (a) LaNiO₃ and (b) LNO-PZT-LNO. Images in a') and b') are the corresponding inclined views.

RESULTS

Figure 1 shows the idea of concentric tubes inside the channels of a porous silicon template after sequential infiltration of LNO, PZT, and LNO precursors. There is also a surface trilayer which could be beneficial for electrical contacting of the tube arrays for the inkjet printing application. However, electrical characterization like the capacitance will be carried out after isolating individual tubes from the rest tubes by focused ion beam.

Figure 2 shows SEM pictures of the LNO and LNO+PZT tube arrays. Note that the fabrication is on wafer scale. The tube wall thickness has a reasonable homogeneity. All the tubes are test-tube like structure (see Fig. 2a) as a replica of the blind pores of the Si templates.

But we have also fabricated 100 μm long tubes with both ends open using the membrane Si templates with see-through channels. The LNO tubes are continuous and robust, which is important for the electrode purpose. The LNO+PZT tubes have a total wall thickness of ca 200 nm and posses their clean hollow interior after four repetitious infiltrations.

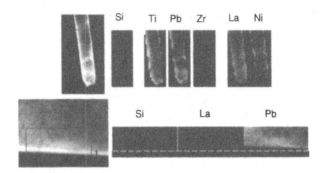

Figure 3 EDX elemental mapping images of the tubes and part of the tube wall. Not shown are the Ni, Zr and Ti. Ni has the same mapping image as La, and Zr/Ti the same as Pb. The horizontal dashed line is a guide for the eye.

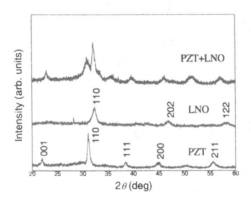

Figure 4 XRD spectra of the PZT, LNO, and LNO+PZT microtube arrays protruding the Si matrix.

Energy dispersive X-ray spectroscopy was used to verify the elements of the tubes. All the elements of LNO and PZT are clearly resolved in the EDX spectrum (data not shown). The EDX elemental mapping images in Fig. 3 show the spatial distribution of the elements. The LNO layer is quite homogeneous whereas the PZT layer is somewhat non-uniform (see top part of Fig. 3).

The mapping images recorded from the edge of a tube (bottom of Fig. 3) shows that Si is on the outside, then La (and Ni) inside in a distinct band, followed by the diffuse PZT inside the LNO. This is consistent with the trilayer structure in the scheme in Fig. 1. It is noted that a Si layer exists on the outer surface of the tube. The Si might be due to diffusion from the template during annealing, as also found in the case of carbon nanotube template (10), but a more likely origin is a SiO_x layer which formed by oxidation and was not removed by KOH.

The XRD spectra of the microtube arrays, measured after partial etching of the Si, are shown in Fig. 4. Clearly all the tubes are crystallized. All the peaks can be indexed to perovskite PZT after the annealing at 650 °C. This is in contrast to Bharadwaja's case, in which a 700 °C annealing temperature was needed to obtain a perovskite phase (8). This could be due to the difference in the precursor types: we used a MOD precursor whereas Bharadwaja et al. used a normal sol precursor. The peaks of LNO tubes are also consistent with literature. The peaks for the LNO+PZT tubes can be assigned according to those of the constituents. The TEM investigations (data not shown here) reveal that the tube wall is polycrystalline containing <50 nm sized crystallites.

The P-E and/or C-V measurement of an individual microtube is underway. Here we show the hysteresis loop of a control sample, which is PZT film of similar thickness to that of the tube wall deposited on a LNO film. The resistivity of the LNO film was estimated to be about two orders higher than Pt based on four-point measurement. The remanent polarization P_r, 15 $\mu C/cm^2$ under an applied voltage of 14 V, is lower than that of PZT film of the same thickness on Pt electrode, 22 $\mu C/cm^2$. But their coercive fields are very close, 200 kV/cm. The rectangularity of hysteresis loop of the PZT/LNO is somewhat worse than that of PZT/Pt. It might be related to the surface roughness of the LNO film due to its poor wettabilty on SiO_2.

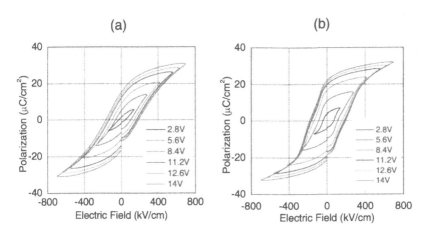

Figure 5 P-E hysteresis loops under different applied voltages for a $\approx$100 nm PZT film deposited on (a) LNO thin film, and (b) Pt coated SiO_2 flat substrates. The top electrodes for both cases are Pt by sputtering through a mesh mask.

CONCLUSIONS

We showed our preliminary results of the fabrication and assembly of concentric LNO-PZT-LNO trilayered microtube arrays. Structural characterization reveals that these tubes are polycrystalline perovskite phase. Mechanical and electrical evaluation of the individual tubes is needed, and currently underway, for their application as inkjet printer head or liquid drug sprayers.

REFERENCES

1. Y. Luo, S.K. Lee, H. Hofmeister, M. Steinhart, U. Gösele, Nano Lett. 4, 143 (2004)
2. M. Steinhart, Z. Jia, A. K. Schaper, R. B. Wehrspohn, U. Gösele, and J. H. Wendorff, Adv. Mater. 15, 706 (2003)
3. B. A. Hernandez, K. S. Chang, E. R. Fisher, and P. K. Dorhout, Chem. Mater. 14, 480 (2002)
4. Y. Luo et al., Appl. Phys. Lett. 83, 440 (2003)
5. F. D. Morrison, L. Ramsay, J. F. Scott, J. Phys.: Condens. Matter 15, L527 (2003).
6. X. Y. Zhang, C. W. Lai, X. Zhao, D. Y. Wang, J. Y. Dai, Appl. Phys. Lett. 87, 143102 (2005)
7. L. Zhao, M. Steinhart, J. Yu, U. Gösele, J. Mater. Res. 21, 685 (2006)
8. S. S. N. Bharadwaja, M. Olszta, S. Trolier-McKinstry, X. Li, T. S. Mayer, F. Roozeboom, J. Am. Ceram. Soc. 89, 2695 (2006)
9. M. Alexe, D. Hesse, V. Schmidt, S. Senz, H. J. Fan, M. Zacharias, U. Gösele, Appl. Phys. Lett. 89, 172907 (2006)
10. S. Kawasaki et al., Appl. Phys. Lett. 92, 053109 (2008)
11. L. Zhao, T. Z. Lu, M. Zacharias, J. Yu, J. Shen, H. Hofmeister, M. Steinhart, U. Gösele, Adv. Mater. 18, 363 (2006)
12. Y. Kim et al., presented in MRS 2008 Spring meeting, San Francisco, in symposium F.
13. K. B. Shelimov, D. N. Davydov, and M. Moskovits, Appl. Phys. Lett. 11, 1722 (2000)
14. X. Y. Zhang, X. Zhao, C. W. Lai, J. Wang, X. G. Tang, and J. Y. Dai, Appl. Phys. Lett. 85, 4190 (2004)
15. X. J. Meng, J. L. Sun, J. Yu, H. J. Ye, S. L. Guo and J. H. Chu, Appl. Surf. Sci. 171, 68 (2001)

Mater. Res. Soc. Symp. Proc. Vol. 1071 © 2008 Materials Research Society   1071-F03-09

Ferroelectric and Electrical Properties of BaZrO3 Doped Sr0.8Bi2.2Ta2O9 Thin Films

Mehmet S. Bozgeyik[1,2], J. S. Cross[3], H. Ishiwara[4], and K. Shinozaki[1]

[1]Metallurgy and Ceramics Science, Tokyo Institute of Technology, 2-12-1 S7 Ookayama, Meguro-ku, Tokyo, 152-8550, Japan
[2]Physics Dept., Kahramanmaras Sutcu Imam Univ., Faculty of Science and Literature, Kahramanmaras, 46100, Turkey
[3]Fujitsu Laboratories, Ltd., 10-1 Morinosato-wakakiya, Atsugi, Kanagawa, 243-0197, Japan
[4]Interdisciplinary Graduate School of Science and Engineering, Tokyo Institute of Technology, 4259 Nagatsuta, Midori-ku, Yokohama, 226-8503, Japan

ABSTRACT

For the first time, $BaZrO_3$ (BZO) doped $Sr_{0.8}Bi_{2.2}Ta_2O_9$ (SBT) thin films were prepared and related ferroelectric and electrical properties were evaluated. Sol-gel thin films of SBT doped with two different BZO mol% ratios were fabricated by spin coating technique on Pt(100nm)/Ti(50nm)/Si(100) substrates. The films were well crystallized at 750 ^{0}C in oxygen gas by RTA for 30 min. From the XRD analysis the crystalline orientation pattern showed that the dominant orientation is (115) for both doped and pure SBT. The films with 5 and 7 mol% BZO ratios showed ferroelectric hysteresis loops at a frequency of 10 kHz. The remanent polarization $(2P_r)$ was significantly reduced to ~5.7 $\mu C/cm^2$ and 1.9 $\mu C/cm^2$ by 5 and 7 mol % doping, respectively. Such a low remanent polarization of 7 mol% BZO doped SBT is suitable for 1T (One Transistor) –FET (Field Effect Transistor) type FeRAMs. The dielectric constant decreased to ~135 by 7 mol% doping compare to that of 205 of SBT. Doping leads to an increase in leakage current to 10^{-7} A/cm^2 level at electric field at 300 kV/cm.

KEYWORDS: Ferroelectric Properties, $Sr_{0.8}Bi_{2.2}Ta_2O_9$, SBT, $BaZrO_3$, Thin Film, FeFET

INTRODUCTION

Ferroelectric Field Effect Transistor (FeFET) 1T-type FeRAM is a particular application in which the stored data can be read out nondestructively and with high device packing density by scaling down. For a FeFET it is desirable to choose a ferroelectric film with a small remanent polarization as well as low dielectric constant, so that the supplied voltage is effectively applied to the ferroelectric film at saturated polarization condition. For ordinary MFIS (Metal-Ferroelectric-Insulator-Si) of FeFETs the charge induced by polarization of ferroelectric film is much larger than the maximum induced charge of a buffer layer at a breakdown voltage; the insulator buffer layer breaks down before the polarization of the ferroelectric film is saturated [1]. Therefore, ferroelectric gate material needs to have low enough remanent polarization to match the charge (~1 $\mu C/cm^2$) required to control the channel conductivity of FET for memory operation. Also, it is a fact that the dielectric constant of typical ferroelectric material is much higher than that of buffer layers, so most of the external voltage is applied to the buffer layer. So, lowering the dielectric constant of ferroelectric material leads to increase the applied voltage in ferroelectric-gate layer in which the ferroelectric hysteresis loop saturates, which is desired to obtain a long data retention time [2], is necessary for memory operation.

SBT is an Aurivillius type oxide, consisting of perovskite layers $(SrTa_2O_7)^{2-}$ sandwiched between $(Bi_2O_2)^{2+}$ layers [3,4]. Although, it is mostly studied for MFIS structure with different buffer layers such as HfO_2, Al_2O_3, SiON, etc. its material parameters is not well match for 1T-type FeRAMs. It has been reported that even small changes in chemical composition, for instance Sr deficiency and/or excess Bi, results in enhanced dielectric and ferroelectric properties including dielectric constant and remanent polarization [5-7]. The objectives in this work are to obtain a lower remanent polarization and dielectric constant based upon SBT for further applications for 1T-type FeRAMs. Our approach is to add BZO, which is cubic perovskite structure with dielectric constant around 36 or less [8-10], into SBT to decrease the remanent polarization and dielectric constant by incorporation larger ionic sizes of Ba and Zr [11] compared to constituent Sr and Ta, respectively. To our knowledge, this is the first time such material properties have been targeted by addition of BZO into SBT. The results indicate this approach is successful to decrease remanent polarization and dielectric constant.

EXPERIMENT

The sol-gel 5 and 7 mol% precursor solution of BZO doped SBT and pure SBT (0.33 mol/L concentration Toshima MFG Co. Ltd.) were spin coated on $Pt/Ti/SiO_2/Si(100)$ substrate at 2500 rpm for 30 sec. The wet films were successively dried on hot plates at 155 ^{0}C for 2 min., 240 ^{0}C for 3 min. and 400 ^{0}C for 2 min to remove the volatile organic compounds, and successively fired in RTA furnace at 750 ^{0}C for 1 min in O_2 atm. This processes were repeated five times to produce film thicknesses of 230 nm, 280 nm and 285 nm for pure, 5 and 7 mol% doped SBTs, respectively. Finally, the films were crystallized at 750 ^{0}C in O_2 atm. for 30 min. Dot-shaped 3.14×10^{-4} and 0.79×10^{-4} cm^{-2} in areas Platinum top electrodes were deposited by Electron Beam Evaporation (E-gun) through a shadow mask for the measurements of electrical and ferroelectric properties. The crystalline structure of the films was investigated with a multipurpose X-ray diffractometer (X`Pert-Pro MPD, Philips). An Atomic Force Microscope of NanoScope III was used to analyze the surface morphology. Ferroelectric hysteresis loops of the thin film were measured at a frequency of 10 kHz at room temperature using a ferroelectric test system (Toyo Corp., FCE-1A/Fop-100V). Leakage current of the films were measured by using HP4156C Precision Semiconductor Parameter Analyzer. Capacitance vs. voltage (C-V) was measured using a LCR Meter (Toyo Corp.) at a frequency of 1 MHz.

DISCUSSION

The x-ray diffraction patterns of pure, 5 and 7 mol% doped SBT films are presented in figure 1, which indicates that all films are polycrystalline with no detectable secondary phase. All the peaks of layered perovskite phase are observed for pure and doped SBT which indicates that the layered perovskite structure was preserved with addition of 7 mol%. By doping the (115) diffraction peak is slightly broaden. This suggests that the orthorhombic structure distortion of doped SBT is reduced. On the other hand, different size of crystalline grains like small crystalline grain size (as shown in figure 4) might cause line broadening [12].

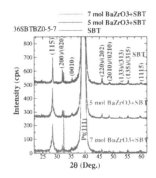

Figure 1. Crystal analysis of pure and BaZrO₃ doped SBT thin films by XRD.

Figure 2 presents the polarization hysteresis loop evaluation at 400 kV/cm electric field.

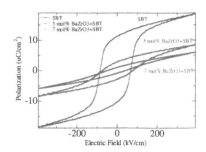

Figure 2. Comparison of hysteresis loops of pure and BaZrO₃ doped SBTs.

Remanent polarization ($2P_r$) is ~21.7 $\mu C/cm^2$ for SBT while for 5 and 7 mol% doped SBTs those are ~5.7 $\mu C/cm^2$ and ~1.9 $\mu C/cm^2$, respectively. The polarizability in isotropic perovskites (ABO_3) is largely determined by the sizes of the A and B cations [13]. The decrease of P_r might be due to decrease the structural distortion of perovskite-type unit by introducing larger ionic radii via decreasing of displacements of corresponding ions in the TaO_6 octahedron (by means of decreasing rattling space) as some different cases discussed in Ref.[14-19].

Pure SBT shows good saturation properties and squareness in figure 3. Doping leads to less saturation and reduction of loop squareness. Formation of small grains is a partial reason for worse saturation as well as little suppression on crystallization by doping.

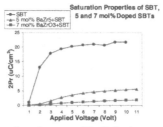

Figure 3. Saturation properties of pure and BaZrO$_3$ doped SBTs.

The film surface morphology analysis in figure 4 showed that grains were densely packed but doping lead to form small grains like ~30 nm in size. Average grain sizes for pure, 5 and 7 mol% doped SBT are 112 nm, 77 nm and 74 nm, respectively. Also, surface roughness (rms) values for pure, 5 and 7 mol% doped SBT are 10.5 nm, 11.9 nm and 18.1 nm, respectively. This makes worse top electrode film interface which causes to increase the leakage current.

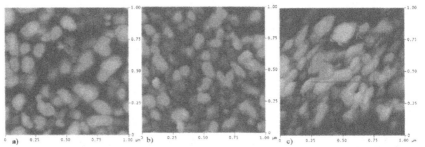

Figure 4. Surface morphology analysis of a) pure SBT, b) 5 mol% and c) 7 mol% BaZrO$_3$ doped SBT films.

Figure 5 indicates the leakage current behaviors. Doping causes to increase the leakage

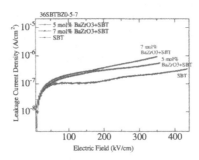

Figure 5. Leakage current behavior of pure and doped SBTs

current density to high10^{-7} A/cm^2 at 300 kV/cm. Currents increase linearly at low applied electric field (the slope of Ln(J) vs. Ln (E) plot (not shown here) is changing between 1-1.5) which indicates Ohmic conduction. At moderate fields up to 200-250 kV/cm Ln(J) vs. E$^{0.5}$ plot is straight which shows the Schottky emission. At higher applied field the current obeys the Frenkel-Poole emission due to that the Ln(J/E) vs. E$^{0.5}$ plot is straight. Higher leakage currents at electric field higher than 300 kV/cm might indeed be associated with increased contribution of trap assisted conduction or hopping mechanism. Increasing the leakage current may also be considered as lowering Schottky barrier height and shallow trap levels and increasing density of traps and defects as well as a lower band gap. In layered structures like SBT the $(Bi_2O_2)^{2+}$ layers mainly control electronic responses like electrical conductivity, band gap, etc. [20, 21], while ferroelectricity mainly arises in the perovskite blocks [22, 23]. Doping of BZO of which the constituent Ba and Zr ion size are larger than the Sr, Bi and Ta, respectively, might lead some stress on $(Bi_2O_2)^{2+}$, which may cause to increase the leakage current. Moreover, that increasing surface roughness causes comparatively worse interface between top electrode and film increases the leakage current. Currently, we are working for further understanding the reasons for increasing leakage current using the films in capacitor and MFIS diode structures as well as changing dopants.

Figure 6 presents the C-V measurements. Butterfly type hysteresis loops of doped SBT

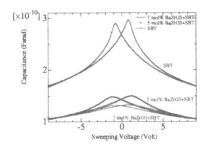

Figure 6. Capacitance (C-V) measurements of pure and BaZrO$_3$ doped SBTs.

once again indicate the ferroelectricity. Capacitance of doped SBT is significantly decreased by doping. Dielectric constants (ε_r), calculated by using zero-volt value of capacitance, of pure, 5 and 7 mol% BZO doped SBTs are around 205, 140 and 135, respectively.

CONCLUSIONS

5 and 7 mol% BZO doped SBT were fabricated by growing thin film on Pt/Ti/SiO$_2$/Si substrate by Sol-gel spin coating technique and investigated through crystal structure and surface morphology analysis as well as measuring ferroelectric and electrical properties further to use as ferroelectric-gate candidate material for 1T-type FeFET memory applications. The (2P$_r$) value was significantly decreased to 1.9 µC/cm^2 by 7 mol% doping. Corresponding dielectric constant successfully decreased to ~135. Doping enhances the leakage current level to high10^{-7} A/cm^2 at 300 kV/cm. The impact of the BZO doping on the SBT crystal structure and its impact on leakage current are under investigation regarding further understanding of the electrical properties as well as different dopants in both capacitor and MFIS structure.

ACKNOWLEDGMENTS

This work was supported by a Grant-in Aid for Scientific Research from the Ministry of Education of Japan, Japan Society for the Promotion of Science (JSPS) and Fujitsu Lab. Ltd.. One of the authors (Mehmet S. Bozgeyik) acknowledges the support from the Scientific and Technological Research Council of Turkey (TUBITAK) for overseas Post-doctoral research.

REFERENCES

1. I. S. -K. Kang and H. Ishiwara: Jpn. J. Appl. Phys. **41** (2002) 2094.
2. E. Tokumitsu, K. Okamoto and H. Ishiwara: Jpn. J. Appl. Phys. **40** (2001) 2917.
3. E. C. Subbarao, J. Phys. Chem. Solids 23, 665(1962). & E. C. Subbarao: J. Am. Ceram. Soc. **45** (1962) 166.
4. Noguchi, M. Miyayama, K. Oikawa, T. Kamiyama, M. Osada, and M. Kakihana, Jpn. J. Appl. Phys. **41** (2002) 7062.
5. H. Watanabe, T. Mihara, H. Yoshimori, and C. A. P. de Araujo: Jpn. J. Appl. Phys., Part1 **34** (1995) 5240.
6. T. Noguchi, T. Hase, and Y. Miyasaka: Jpn. J. Appl. Phys., Part1 **35** (1996) 4900.
7. T. Atsuki, N. Soyama, T. Yonezawa, and K. Ogi: Jpn. J. Appl. Phys., Part1 **34** (1995) 5096.
8. I. Levin, T. G. Amos, S. M. Bell, L. Farber, T. A. Vanderah, R. S. Roth and B. H. Toby: J. Solid. State Chem. **175** (2003) 170.
9. T. Maiti, R. Guo and A. S. Bhalla: J. Appl. Phys. **100** (2006) 114109.
10. C.-L. Hung, Y. –L. Chueh, T. –B. Wu and L. –J. Chou: J. Appl. Phys. **97** (2005) 034105.
11. J. W. Bennett, I. Grinberg and A. M. Rappe: Phys. Rev. B **73** (2006) 180102(R).
12. H.-Y. Chou, T. –M. Chen and T.-Y. Tseng: Mater. Chem. Phys. **82** (2003) 826.
13. F. Jona and G. Shirane: *Ferroelectric crystals* (Pergamon Press, New York, 1962).
14. M. Q. Cai, Z. Yin, M. S Zhang, Solid State Commun.: **133** (2005) 663.
15. Y. Shimakawa, Y. Kubo, Nakagawa, S. Goto, T. Kamiyama, H. Asano and F. Izumi: Phys. Rev. B, **61** (2000) 6560.
16. Y. Shimakawa, Y. Kubo, Y. Tauchi, T. Kamiyama, H. Asano and F. Izumi: Appl. Phys. Let. **77** (2000) 2749.
17. P. Goel and K. L. Yadav: Physica B **382** (2006) 245.
18. H. Irie, M. Miyayama and T. Kudo: J. Appl. Phys. **90** (2001) 4089.
19. M. M. Kumar and Z.-G. Ye: J. Appl. Phys. **90** (2001) 934.
20. S. K. Kim, M. Miyayama, and H. Yanagida, Mater. Res. Bull. **31** (1996) 121.
21. M. Stachiotti, C. Rodriguez, C. A. Draxl, and N. Christensen: Phys. Rev. B **61** (2000) 14434.
22. Y. Shimakawa, Y. Kudo, Y. Nakagawa, T. Kamiyama, H. Asano, and F. Izumi: Appl. Phys. Lett. **74** (1999) 1904.
23. Y. Noguchi, M. Miyayama, and T. Kudo: Phys. Rev. B **63** (2001) 214102.

Mater. Res. Soc. Symp. Proc. Vol. 1071 © 2008 Materials Research Society   1071-F03-14

Stability of Larger Ferromagnetic Chain-of-sphere Nanostructure Comprising Magnetic Vortices

Prabeer Barpanda
Malcolm G McLaren Centre for Ceramic Research, Department of Materials Science and
Engineering, Rutgers University, 607, Taylor Road, Busch Campus, Piscataway, NJ, 08854-8065

ABSTRACT

Chain-of-sphere (CoS) nanostructure containing Permalloy ($Fe_{20}Ni_{80}$) nanospheres of uniform size (d=50 nm) has been studied using micromagnetic simulation. These large-size Permalloy nanospheres support magnetic vortex structure upon relaxation. The presence of magnetic vortices in CoS architecture affects its magnetic properties significantly. Micromagnetic behaviour of Permalloy CoS system was studied focusing on the magnetization reversal process. The presence of magnetic vortices triggers a vortex creation and annihilation mechanism (VCA) involving the formation and breaking of an inversion symmetry (IS) feature. This VCA mechanism has been studied using 3D micromagnetic simulation and results of coercivity and vortex parameters are presented.

INTRODUCTION

Magnetic nanostructures (nanodots, nanorings, nanobars, nanocones) are subjected to wide-scale research to explore their potential use in applications ranging from magnetic sensors to high-density magnetic recording [1,2,3]. Magnetic properties of ferromagnetic nanostructures are mainly controlled by the shape, dimensions and material parameters. Thus, research on nanostructure has been extended to 1D (nanobars), 2D (dots, thin-films) and 3D (truncated cones etc) systems with different shape/ size. One less explored nanostructure is the ferromagnetic chain-of-sphere (CoS) structure, originally proposed by Bean et al [4,5] to explain the magnetization reversal behaviour of elongated spherical particles. The CoS system can serve as a suitable model for understanding the magnetic phenomena of elongated structures like magnetic nanowires, ellipsoids etc. Permalloy CoS with large sphere size (d> 25 nm) supports magnetic vortices in individual sphere [6]. These vortices have been observed by off-axis electron holography in Fe_xNi_{1-x} CoS produced by vapour phase condensation route [7, 8].

The current work is an attempt to study the ferromagnetic chain-of-sphere nanostructure focusing on its underlying magnetization reversal process involving bi-domain magnetic vortex. This modeling study can gain basic insight for designing magnetic devices with thin nanowires and ellipsoids. As observed earlier, the chain comprising larger spheres with magnetic vortices reverses gradually involving vortex creation and annihilation (VCA) mechanism [6]. Here using micromagnetic simulation, Permalloy CoS system having 5 uniform spheres (size, d = 50 nm) has been examined, focusing on the magnetization reversal process. The magnetization reversal mechanism in chain structure has been studied as a function of number of sphere. The variation in coercivity, remanence, vortex parameters (vortex core diameter and vortex rotation plane) and exchange energy has been discussed for the Permalloy chain-of-sphere system.

SIMULATION

Micromagnetic simulation [9] is extensively used to study the physics of magnetism. Basically, it assumes any magnetic system to be an integration of large number of individual, randomly aligned magnetic spins. Then, it solves the Landau-Lifshitz-Gilbert (LLG) equation, expressed as:

$$(1+\alpha^2) \, \delta M / \, \delta t = -\gamma(M \times \mu_0 H_{eff}) - (\alpha\gamma/M_s) \, M \times (M \times \mu_0 H_{eff})$$

where, α is damping parameter and γ is gyromagnetic ratio of individual electron spin.

A linear CoS structure (size 50 nm) is shown in figure 1 (axis along x-direction). Each sphere is spatially divided into cubic cells ($N_x \times N_y \times N_z$) of uniform magnetization, with demagnetization field computed to all orders. Micromagnetic simulations were performed using a 3D-FFT micromagnetic simulator [10]. To ensure the accuracy of simulation results, these cubic cells were made smaller than the characteristic exchange length of Permalloy $l_{ex} = [2A_{ex}/ (\mu_0 M_s^2)]^{1/2} = 5.1$ nm. Here, the equilibrium criterion was set when the maximum change in torque is below a given tolerance of m $\times H_{eff} <= 10^{-4} M_s$. A gyromagnetic frequency γ of 17.6 and a damping parameter α value of 1 were used throughout to capture the equilibrium static configurations with good computational speed. For Permalloy, saturation magnetization (M_s) of 800 emu/cc, exchange stiffness (A) of 1.05 µerg/cm and uniaxial anisotropy (K_u) of 1000 erg/cc were used. Magnetization reversal mechanism was studied by applying axial external field from -4 ~ 4 kOe and by changing the number of sphere from 2 to 9. Thermal fluctuations were not considered. The simulations are independent of anisotropy direction and initial magnetization.

DISCUSSION

Magnetization reversal involving vortices

During the hysteresis cycle, magnetization reversal occurs with vortex creation and annihilation (VCA) mechanism (Fig 2a-e). The CoS system is initially saturated along the field direction (a). Upon lowering the field, with magnetic relaxation in CoS, magnetic vortices appear in the end spheres (b). However, these vortices were found to swirl in opposite direction. With further lowering of external field, magnetic vortices enter middle spheres with the centre sphere still in saturated state (c). Permalloy sphere of 50 nm size naturally supports magnetic vortices. Thus, with gradual lowering of external field, twisted magnetic vortices appear in spheres, which grow to become complete vortices with a vertical saturated domain in center part surrounded by a horizontal circular swirling domain. When the external field is decreased further, gradually the magnetic vortices get deformed by swirling in the opposite direction. In all the spheres, the outer swirling domain gradually switches to opposite direction followed by instant domain switching of central saturated domain. It was observed the centre sphere never get a chance to develop magnetic vortex. But the reversal of adjacent spheres forces the centre sphere to reverse like a single domain. The magnetic vortices always enter the CoS from both ends of the chain.

This VCA mechanism hold good for longer chain with more number of spheres. Figure 3 shows a series of CoS system with different number of sphere. In each case, midway during reversal, half of the structure is found to have vortices swirling anticlockwise (left-handed), while the other half vortices swirling clockwise (right-handed).

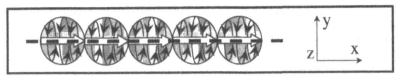

Figure 1: The geometry of chain-of-sphere system with chain axis parallel to x-axis.

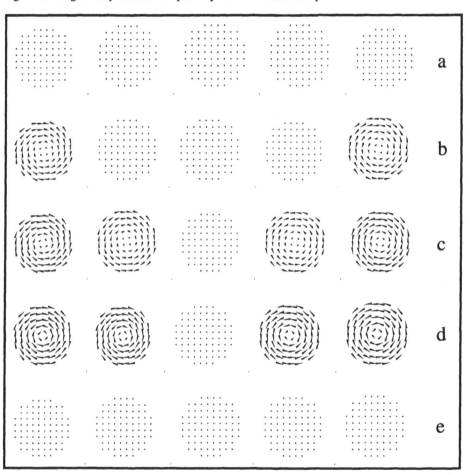

Figure 2: Micromagnetic snapshots demonstrating magnetization reversal mechanism (external field applied: -4000 ~ 4000 Oe). The vortex creation and annihilation mechanism and inversion symmetry feature (step c and d) in a Permalloy chain of 5 spheres (d = 50 nm) is shown. For convenience, the central y-z cross-section of each sphere (in CoS) is shown to capture magnetic vortices during reversal process.

Due to the equal number of oppositely aligned magnetic vortices, the CoS system shows an inversion symmetry feature during the reversal (fig 2c,2d). This type of feature can be observed in wider nanobars, ellipsoids etc. The inversion symmetry feature can favor minimum Gibb's free energy in the chain structure. During the overall magnetization reversal, the inversion symmetry feature appears and disappears midway.

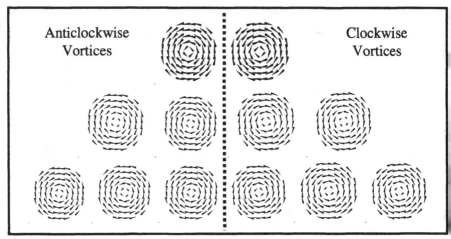

Figure 3: Micromagnetic snapshots during the midway of reversal process showing the inversion symmetry feature in Permalloy chain structure for different number of sphere (n=2,4,6). While the left half of any structure is swirling anticlockwise, the right half is swirling clockwise, thus creating an inversion symmetry feature with centre at the physical centre of whole structure. For each sphere in CoS, the central y-z cross-section is shown to capture magnetic vortex.

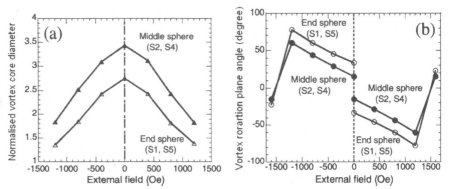

Figure 4: The variation of (a) vortex core diameter (VCD) and (b) vortex rotation plane (VRP) of each sphere in a Permalloy CoS structure with 5 spheres (d = 50 nm) (the external field applied: -2000 ~ 2000 Oe) during a hysteresis cycle.

Vortex parameters (vortex core diameter and vortex rotation plane) of CoS

As described earlier, the chain of larger sphere involves creation, growth and annihilation of magnetic vortices. These vortices are affected by the external field and interacting adjacent spheres. As a result, the vortex core diameter (VCD i.e. central saturated domain part) and the vortex rotation plane (VRP i.e. angle of swirling moments with y-z plane) gradually changes during hysteresis cycle. Figure 4 summarizes these vortex parameters for all spheres in a CoS system. The VCD decreases from 0 Oe on both side, indicating the formation of more stable vortices i.e. with a very confined central saturated domain surrounded by swirling moments in majority part of the sphere. As the vortices appear in the end sphere first, end spheres possess more perfect vortices throughout the hysteresis cycle. Thus, VCD values are less for end spheres than intermediate spheres (Fig 4a). The VRP variation (Fig 4b) shows steady growth of magnetic vortices with magnetization reversal and vortex switching around ±1200 Oe marked by abrupt change in VRP.

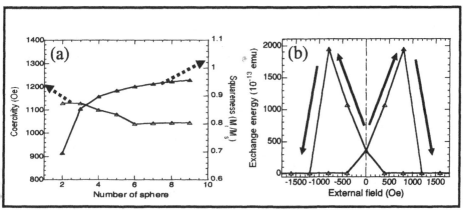

Figure 5: (a) Variation of coercivity and remanence in Permalloy CoS system of uniform sphere size (d=50 nm) for different number of spheres comprising the chain. (b) A typical variation of exchange energy during hysteresis cycle. The axial external field is varied from -4000~4000 Oe.

Coercivity and remanence of CoS

Smaller spheres with single domain need very high external field for magnetization reversal via domain switching, thus giving rise to high coercivity. However, larger sphere with vortices undergo gradual reversal with VCA mechanism, which lowers the coercivity. When the number of sphere in the CoS structure increases, the vortex creation and propagation throughout chain (from both ends) becomes easier due to exchange interaction between adjacent spheres. As a result, the overall coercivity value decreases as shown in figure 5a. It is marked with increase in the number of sphere from 2 to 9, the coercivity value decreases gradually from 1128 to 1045 Oe. The squareness value (remanent magnetization/ saturation magnetization i.e. Mr/Ms) increases slightly from 0.7 to 0.95 with longer CoS structure (Fig 5a). A typical variation of exchange energy in permalloy chain during hysteresis cycle is shown in figure 5b.

CONCLUSIONS

Magnetic property of linear chain-of-sphere nanostructure (uniform sphere size, 50 nm) was studied using micromagnetic simulation. The presence of magnetic vortices in 50 nm wide Permalloy sphere favors the evolution, growth and annihilation of vortices in individual sphere during magnetization reversal. This vortex creation/ annihilation mechanism and appearance of inversion symmetry feature have been captured using simulation. This gradual reversal mechanism was correlated to the relatively lower value of coercivity. The vortex parameters (VCD and VRP) and exchange energy were calculated using micromagnetic simulation. The CoS forms an interesting nanomagnetic system for studying elongated structure like ferromagnetic nanowires. A detail study of effect of aspect ratio and incidence angle of external field on magnetic properties of CoS system is presented elsewhere [11].

ACKNOWLEDGMENTS

The author wish to thank Dr. T. Kasama (Cambridge-UK), Prof. M.R. Scheinfein (Simon Fraser University-Canada) and Prof. R.E. Dunin-Borkowski (TU-Denmark) for their technical help on chain-of-sphere problem. The author thanks Dr. Anil Kaza (Intel, OR), Dr. Sai K Doddi, Dr. Pradeep George and Prof. Glenn G Amatucci (Rutgers, NJ) for their kind support throughout the work.

REFERENCES

1. C.A. Ross, Ann. Rev. Mater. Sci. 10(3) (1998) 247-264.
2. R.P. Cowburn, J. Phys D: Appl. Phys. 33(1) (2000) R1-R16.
3. C.A. Ross, M. Farhoud, M. Hwang et al., J. Appl. Phys., 89 (2001) 1310-1319.
4. I.S. Jacobs, C.P. Bean, Phys. Rev. 100 (1955) 1060-1067.
5. C.P. Bean, J. Livingston, J. Appl. Phys., 30 (1959), 120S-129S.
6. P. Barpanda, T. Kasama et al, J. Mag Mag Mater, Submitted (2006).
7. M.J. Hytch, R.E. Dunin-Borkowski et al, Phys. Rev. Lett. 91 (2003), 257207.
8. R.K.K. Chong, R.E. Dunin-Borkowski et al, Inst. Phys. Conf. Ser. 179 (2003) 451-454.
9. W. F. Brown Jr, Miromagnetics, Interscience, New york (1963).
10. http://llgmicro.mindspring.com
11. P. Barpanda, Comp. Mater. Sci. (2008), Submitted.

AUTHOR INDEX

SUBJECT INDEX

221

Printed in the United States
By Bookmasters